D1564216

COMMUTATIVE RINGS WITH ZERO DIVISORS

MONOGRAPHS AND TEXTBOOKS IN PURE AND APPLIED MATHEMATICS

1. *K. Yano,* Integral Formulas in Riemannian Geometry (1970)*(out of print)*
2. *S. Kobayashi,* Hyperbolic Manifolds and Holomorphic Mappings (1970) *(out of print)*
3. *V. S. Vladimirov,* Equations of Mathematical Physics (A. Jeffrey, editor; A. Littlewood, translator) (1970) *(out of print)*
4. *B. N. Pshenichnyi,* Necessary Conditions for an Extremum (L. Neustadt, translation editor; K. Makowski, translator) (1971)
5. *L. Narici, E. Beckenstein, and G. Bachman,* Functional Analysis and Valuation Theory (1971)
6. *S. S. Passman,* Infinite Group Rings (1971)
7. *L. Dornhoff,* Group Representation Theory (in two parts). Part A: Ordinary Representation Theory. Part B: Modular Representation Theory (1971, 1972)
8. *W. Boothby and G. L. Weiss (eds.),* Symmetric Spaces: Short Courses Presented at Washington University (1972)
9. *Y. Matsushima,* Differentiable Manifolds (E. T. Kobayashi, translator) (1972)
10. *L. E. Ward, Jr.,* Topology: An Outline for a First Course (1972) *(out of print)*
11. *A. Babakhanian,* Cohomological Methods in Group Theory (1972)
12. *R. Gilmer,* Multiplicative Ideal Theory (1972)
13. *J. Yeh,* Stochastic Processes and the Wiener Integral (1973) *(out of print)*
14. *J. Barros-Neto,* Introduction to the Theory of Distributions (1973) *(out of print)*
15. *R. Larsen,* Functional Analysis: An Introduction (1973) *(out of print)*
16. *K. Yano and S. Ishihara,* Tangent and Cotangent Bundles: Differential Geometry (1973) *(out of print)*
17. *C. Procesi,* Rings with Polynomial Identities (1973)
18. *R. Hermann,* Geometry, Physics, and Systems (1973)
19. *N. R. Wallach,* Harmonic Analysis on Homogeneous Spaces (1973) *(out of print)*
20. *J. Dieudonné,* Introduction to the Theory of Formal Groups (1973)
21. *I. Vaisman,* Cohomology and Differential Forms (1973)
22. *B. -Y. Chen,* Geometry of Submanifolds (1973)
23. *M. Marcus,* Finite Dimensional Multilinear Algebra (in two parts) (1973, 1975)
24. *R. Larsen,* Banach Algebras: An Introduction (1973)
25. *R. O. Kujala and A. L. Vitter (eds.),* Value Distribution Theory: Part A; Part B: Deficit and Bezout Estimates by Wilhelm Stoll (1973)
26. *K. B. Stolarsky,* Algebraic Numbers and Diophantine Approximation (1974)
27. *A. R. Magid,* The Separable Galois Theory of Commutative Rings (1974)
28. *B. R. McDonald,* Finite Rings with Identity (1974)
29. *J. Satake,* Linear Algebra (S. Koh, T. A. Akiba, and S. Ihara, translators) (1975)

30. *J. S. Golan,* Localization of Noncommutative Rings (1975)
31. *G. Klambauer,* Mathematical Analysis (1975)
32. *M. K. Agoston,* Algebraic Topology: A First Course (1976)
33. *K. R. Goodearl,* Ring Theory: Nonsingular Rings and Modules (1976)
34. *L. E. Mansfield,* Linear Algebra with Geometric Applications: Selected Topics (1976)
35. *N. J. Pullman,* Matrix Theory and Its Applications (1976)
36. *B. R. McDonald,* Geometric Algebra Over Local Rings (1976)
37. *C. W. Groetsch,* Generalized Inverses of Linear Operators: Representation and Approximation (1977)
38. *J. E. Kuczkowski and J. L. Gersting,* Abstract Algebra: A First Look (1977)
39. *C. O. Christenson and W. L. Voxman,* Aspects of Topology (1977)
40. *M. Nagata,* Field Theory (1977)
41. *R. L. Long,* Algebraic Number Theory (1977)
42. *W. F. Pfeffer,* Integrals and Measures (1977)
43. *R. L. Wheeden and A. Zygmund,* Measure and Integral: An Introduction to Real Analysis (1977)
44. *J. H. Curtiss,* Introduction to Functions of a Complex Variable (1978)
45. *K. Hrbacek and T. Jech,* Introduction to Set Theory (1978)
46. *W. S. Massey,* Homology and Cohomology Theory (1978)
47. *M. Marcus,* Introduction to Modern Algebra (1978)
48. *E. C. Young,* Vector and Tensor Analysis (1978)
49. *S. B. Nadler, Jr.,* Hyperspaces of Sets (1978)
50. *S. K. Segal,* Topics in Group Rings (1978)
51. *A. C. M. van Rooij,* Non-Archimedean Functional Analysis (1978)
52. *L. Corwin and R. Szczarba,* Calculus in Vector Spaces (1979)
53. *C. Sadosky,* Interpolation of Operators and Singular Integrals: An Introduction to Harmonic Analysis (1979)
54. *J. Cronin,* Differential Equations: Introduction and Quantitative Theory (1980)
55. *C. W. Groetsch,* Elements of Applicable Functional Analysis (1980)
56. *I. Vaisman,* Foundations of Three-Dimensional Euclidean Geometry (1980)
57. *H. I. Freedman,* Deterministic Mathematical Models in Population Ecology (1980)
58. *S. B. Chae,* Lebesgue Integration (1980)
59. *C. S. Rees, S. M. Shah, and C. V. Stanojević,* Theory and Applications of Fourier Analysis (1981)
60. *L. Nachbin,* Introduction to Functional Analysis: Banach Spaces and Differential Calculus (R. M. Aron, translator) (1981)
61. *G. Orzech and M. Orzech,* Plane Algebraic Curves: An Introduction Via Valuations (1981)
62. *R. Johnsonbaugh and W. E. Pfaffenberger,* Foundations of Mathematical Analysis (1981)
63. *W. L. Voxman and R. H. Goetschel,* Advanced Calculus: An Introduction to Modern Analysis (1981)
64. *L. J. Corwin and R. H. Szcarba,* Multivariable Calculus (1982)
65. *V. I. Istrătescu,* Introduction to Linear Operator Theory (1981)
66. *R. D. Järvinen,* Finite and Infinite Dimensional Linear Spaces: A Comparative Study in Algebraic and Analytic Settings (1981)

67. *J. K. Beem and P. E. Ehrlich,* Global Lorentzian Geometry (1981)
68. *D. L. Armacost,* The Structure of Locally Compact Abelian Groups (1981)
69. *J. W. Brewer and M. K. Smith, eds.,* Emmy Noether: A Tribute to Her Life and Work (1981)
70. *K. H. Kim,* Boolean Matrix Theory and Applications (1982)
71. *T. W. Wieting,* The Mathematical Theory of Chromatic Plane Ornaments (1982)
72. *D. B. Gauld,* Differential Topology: An Introduction (1982)
73. *R. L. Faber,* Foundations of Euclidean and Non-Euclidean Geometry (1983)
74. *M. Carmeli,* Statistical Theory and Random Matrices (1983)
75. *J. H. Carruth, J. A. Hildebrant, and R. J. Koch,* The Theory of Topological Semigroups (1983)
76. *R. L. Faber,* Differential Geometry and Relativity Theory: An Introduction (1983)
77. *S. Barnett,* Polynomials and Linear Control Systems (1983)
78. *G. Karpilovsky,* Commutative Group Algebras (1983)
79. *F. Van Oystaeyen and A. Verschoren,* Relative Invariants of Rings: The Commutative Theory (1983)
80. *I. Vaisman,* A First Course in Differential Geometry (1984)
81. *G. W. Swan,* Applications of Optimal Control Theory in Biomedicine (1984)
82. *T. Petrie and J. D. Randall,* Transformation Groups on Manifolds (1984)
83. *K. Goebel and S. Reich,* Uniform Convexity, Hyperbolic Geometry, and Nonexpansive Mappings (1984)
84. *T. Albu and C. Năstăsescu,* Relative Finiteness in Module Theory (1984)
85. *K. Hrbacek and T. Jech,* Introduction to Set Theory, Second Edition, Revised and Expanded (1984)
86. *F. Van Oystaeyen and A. Verschoren,* Relative Invariants of Rings: The Noncommutative Theory (1984)
87. *B. R. McDonald,* Linear Algebra Over Commutative Rings (1984)
88. *M. Namba,* Geometry of Projective Algebraic Curves (1984)
89. *G. F. Webb,* Theory of Nonlinear Age-Dependent Population Dynamics (1985)
90. *M. R. Bremner, R. V. Moody, and J. Patera,* Tables of Dominant Weight Multiplicities for Representations of Simple Lie Algebras (1985)
91. *A. E. Fekete,* Real Linear Algebra (1985)
92. *S. B. Chae,* Holomorphy and Calculus in Normed Spaces (1985)
93. *A. J. Jerri,* Introduction to Integral Equations with Applications (1985)
94. *G. Karpilovsky,* Projective Representations of Finite Groups (1985)
95. *L. Narici and E. Beckenstein,* Topological Vector Spaces (1985)
96. *J. Weeks,* The Shape of Space: How to Visualize Surfaces and Three-Dimensional Manifolds (1985)
97. *P. R. Gribik and K. O. Kortanek,* Extremal Methods of Operations Research (1985)
98. *J.-A. Chao and W. A. Woyczynski, eds.,* Probability Theory and Harmonic Analysis (1986)
99. *G. D. Crown, M. H. Fenrick, and R. J. Valenza,* Abstract Algebra (1986)
100. *J. H. Carruth, J. A. Hildebrant, and R. J. Koch,* The Theory of Topological Semigroups, Volume 2 (1986)

101. *R. S. Doran and V. A. Belfi,* Characterizations of C*-Algebras: The Gelfand-Naimark Theorems (1986)
102. *M. W. Jeter,* Mathematical Programming: An Introduction to Optimization (1986)
103. *M. Altman,* A Unified Theory of Nonlinear Operator and Evolution Equations with Applications: A New Approach to Nonlinear Partial Differential Equations (1986)
104. *A. Verschoren,* Relative Invariants of Sheaves (1987)
105. *R. A. Usmani,* Applied Linear Algebra (1987)
106. *P. Blass and J. Lang,* Zariski Surfaces and Differential Equations in Characteristic $p > 0$ (1987)
107. *J. A. Reneke, R. E. Fennell, and R. B. Minton.* Structured Hereditary Systems (1987)
108. *H. Busemann and B. B. Phadke,* Spaces with Distinguished Geodesics (1987)
109. *R. Harte,* Invertibility and Singularity for Bounded Linear Operators (1988).
110. *G. S. Ladde, V. Lakshmikantham, and B. G. Zhang,* Oscillation Theory of Differential Equations with Deviating Arguments (1987)
111. *L. Dudkin, I. Rabinovich, and I. Vakhutinsky,* Iterative Aggregation Theory: Mathematical Methods of Coordinating Detailed and Aggregate Problems in Large Control Systems (1987)
112. *T. Okubo,* Differential Geometry (1987)
113. *D. L. Stancl and M. L. Stancl,* Real Analysis with Point-Set Topology (1987)
114. *T. C. Gard,* Introduction to Stochastic Differential Equations (1988)
115. *S. S. Abhyankar,* Enumerative Combinatorics of Young Tableaux (1988)
116. *H. Strade and R. Farnsteiner,* Modular Lie Algebras and Their Representations (1988)
117. *J. A. Huckaba,* Commutative Rings with Zero Divisors (1988)
118. *W. D. Wallis,* Combinatorial Designs (1988)
119. *W. Więsław,* Topological Fields (1988)

Other Volumes in Preparation

COMMUTATIVE RINGS WITH ZERO DIVISORS

James A. Huckaba
University of Missouri—Columbia
Columbia, Missouri

MARCEL DEKKER, INC. New York and Basel

Library of Congress Cataloging-in-Publication Data

Huckaba, James A.
Commutative rings with zero divisors / James A. Huckaba.
p. cm. — (Monographs and textbooks in pure and applied mathematics ; v. 117)
Bibliography: p.
Includes index.
ISBN 0-8247-7844-8
1. Commutative rings. 2. Divisor theory. I. Title. II. Series.
QA251.3.H83 1988
512′.4—dc19 87-36864
CIP

MARCEL DEKKER, INC.
270 Madison Avenue, New York, New York 10016

Current printing (last digit):
10 9 8 7 6 5 4 3 2 1

PRINTED IN THE UNITED STATES OF AMERICA

TO
BEVERLY

Preface

The ideal theoretic method for studying commutative rings has a long and fruitful history. For many aspects of ring theory (primary decomposition, dimension theory, quotient rings, the principal ideal theorem) the use of the method does not depend on the underlying rings being integral domains. In other areas of commutative rings, the ideal theoretic method has been useful in studying integral domains, but until recently has not been used to study rings with zero divisors. Much of multiplicative ideal theory falls into this category. In still other areas there are important questions concerning the ideals of rings with zero divisors that do not have counterparts for integral domains. An example of this is the problem of determining when the space of minimal prime ideals of a commutative ring is compact.

In 1957 Pierre Samuel published a paper, *La notion de place dans un anneau*, in which he gave several extensions of valuation theory to arbitrary commutative rings. Dating from this paper, a considerable amount of research has been devoted to extending multiplicative ideal theory to rings containing zero divisors. This work falls into two parts. First, there are the results that carry over, with little change, from domains to arbitrary commutative rings. (This is not to say that the proofs of these theorems are necessarily straightforward generalizations of the proofs for domains; many times quite the opposite is true.) Second are the results that have strikingly different forms than the corresponding results for integral domains. This class of results gives new insights into the structure of rings.

I have for some time wanted to give an account of commutative ring theory, for rings containing zero divisors, by ideal theoretic methods. This book is my attempt to do so. The topics that are included in this volume are certainly influenced by my prejudices. I have, however, tried to include those topics about zero divisors that will give the reader a coherent picture of the research that has been done on these rings over the past three decades.

There is little overlap of the material presented in this book with the material in other books on commutative algebra. The only way to avoid the overlap is to quote freely from other treatises on the subject, which I have done. Hence, there is a section at the end of the text entitled References. This contains a complete list of commutative algebra books that are referenced in the text. The references are given as in the following example: "By Nakayama's lemma [K, p. 51]...."

There is also a bibliography at the end of the book that lists the resources used in writing this manuscript. References from the bibliography appear only in the Notes section at the end of each chapter. While not all of the items in the bibliography found their way into these sections, all were useful in preparing the manuscript. I have used the Notes for several different purposes: (1) to give historical comments on the chapter immediately preceding it; (2) to give credit for the major results of the text; (3) to explain why a particular point of view was chosen from several competing points of view; (4) to point out alternative terminology and notation that is in the literature; (5) to give the reader references for topics related to, but not covered in, the text.

Since this is basically a book about rings and their ideals, I have used ideal theoretic arguments whenever possible. Hence many of the proofs from the literature have been reworked. However, I did not hesitate to use other methods when no ideal theoretic proof was available.

I was fortunate to have several of my colleagues agree to critically read this manuscript. Dan Anderson, Fred Call, Sam Huckaba, Tom Lucas, and Ira Papick all read sections of the book and gave me valuable suggestions. I wish to thank each of them for their help.

I am especially indebted to two people: Beverly Huckaba and Dan Anderson. My wife, Beverly, not only typed the first two versions of this book, but encouraged me (always with patience and good humor) throughout the entire project. Dan Anderson read more than half the manuscript. With his permission, several of his unpublished results appear in these pages for the first time.

James A. Huckaba

Contents

Chapter I

Total Quotient Rings

1. Preliminaries

All rings considered in this book are assumed to be commutative and have an identity $\neq 0$. A ring R is *reduced* if its nilradical is zero. Elements of R that are not zero divisors are called *regular.* A *regular ideal* of R is one that contains a regular element.

NOTATION. Let R be a ring.

$T(R)$	=	total quotient ring of R
$Z(R)$	=	set of zero divisors of R
$N(R)$	=	nilradical of R
R'	=	integral closure of R in its total quotient ring $T(R)$, or sometimes in an extension ring S of R
$\operatorname{Ann} H$	=	annihilator of a subset H of R; or the annihilator of a subset H of an R-module M

Spec R	=	set of all prime ideals of R; sometimes, but not always, endowed with the Zariski topology.
Max R	=	maximal ideals of R
Min R	=	minimal prime ideals of R
$I[X]$	=	$IR[X]$, where I is an ideal of R and $R[X]$ is the polynomial ring over R
$\subseteq$	=	containment
$\subset$	=	proper containment
A_v	=	$(A^{-1})^{-1}$, for a fractional ideal of R

An *overring* of R is a ring between R and $T(R)$. If R is a subring of a ring S where $S \not\subseteq T(R)$, then S is called an *extension ring* of R. Assume that extension rings have the same identity as R, unless otherwise stated. When no ambiguity may arise we sometimes write $R = S$ for $R \cong S$. In addition, we use the standard notation as conveyed in [G] and [K].

2. Property A

Let R be a commutative ring with total quotient ring $T(R)$. If S is a overring of R, then the map $P \mapsto PT(R) \cap S$ is a one-to-one order preserving correspondence between the prime ideals of R contained in $Z(R)$ and the prime ideals of S contained in $Z(S)$.

Denote the set of minimal prime ideals of R by Min R; for example, if R is an integral domain, then Min R consists only of the zero ideal. The minimal prime ideals of R and the minimal prime ideals of an arbitrary ideal I are of special interest. Useful properties of these primes are contained in the first five results.

Theorem 2.1 Let $P \supseteq I$ be ideals of a ring R, where P is a prime ideal. Then the following conditions are equivalent:

(1) P is a minimal prime ideal of I.

(2) $R \setminus P$ is a multiplicatively closed set that is maximal with respect to missing I.

(3) For each $x \in P$, there is a $y \notin P$ and a nonnegative integer i such that $yx^i \in I$.

PROOF: (1) $\Rightarrow$ (2): Expand $R \setminus P$ to a multiplicatively closed set S that is maximal with respect to missing I. If Q is an ideal containing I that is maximal with respect to being disjoint from S, then Q is prime [K, p.1]. Note that Q is disjoint from $R \setminus P$ which implies that $Q = P$. Thus $R \setminus P = S$.

(2) $\Rightarrow$ (3): Choose a nonzero x in P and let $S = \{yx^i : y \in R \setminus P,\ i = 0, 1, 2, \ldots\}$. Then S is a multiplicatively closed set that properly contains $R \setminus P$. So there is some $y \in R \setminus P$ and a nonnegative integer i such that $yx^i \in I$.

(3) $\Rightarrow$ (1): Assume that $I \subset Q \subseteq P$, where Q is a prime ideal. If there exists some $x \in P \setminus Q$, then there is a $y \notin P$ and a positive integer i such that $yx^i \in I \subset Q$, a contradiction. Therefore $P = Q$.

Corollary 2.2 If R is a reduced ring and if P is a prime ideal of R, then P is a minimal prime ideal of R if and only if for each x in P there exists some $y \notin P$ such that $xy = 0$.

Corollary 2.3 Let J be a finitely generated ideal of a reduced ring R. Then J is contained in a minimal prime ideal of R if and only if $\operatorname{Ann} J \neq 0$.

PROOF: Let $J = (a_1, \ldots, a_n)$. If J is contained in a minimal prime ideal P of R, then there exist $u_i \in R \setminus P$ such that $u_i a_i = 0$ for $i = 1, \ldots, n$. Then $u = u_1 u_2 \cdots u_n$ is a nonzero element belonging to $\operatorname{Ann} J$.

Conversely, if $\operatorname{Ann} J \neq 0$, then $\operatorname{Ann} J \not\subseteq P$ for an appropriate minimal prime ideal P of R. Thus $J \subseteq P$.

Corollary 2.4 In a reduced ring R, $Z(R) = \cup P$, where P varies over $\operatorname{Min} R$.

PROOF: If $x \in Z(R)$, then $\operatorname{Ann}(x) \neq 0$. Hence $x \in P$, for some $P \in \operatorname{Min} R$.

Theorem 2.5 Let R be a ring and let $\{P_\alpha\} = \operatorname{Min} R$. Then $\operatorname{Min} R$ is finite if and only if $P_\alpha \not\subseteq \cup_{\beta\neq\alpha} P_\beta$, for each α.

PROOF: ($\Rightarrow$): This direction is well known.

($\Leftarrow$): Without loss of generality, assume that R is a reduced ring. By using the definition of localization along with Corollary 2.2, it is easy to see that R_P is a field for each minimal prime ideal P of R. Let A be the direct product $\prod R_{P_\alpha}$ of the R_{P_α}'s. Embed R into A via the map $r \mapsto \{\theta_\alpha(r)\}$, where θ_α is the canonical homomorphism of R into R_{P_α}. Therefore R can be thought of as a subring of A.

If $\operatorname{Min} R$ is infinite, then the direct sum $\sum R_{P_\alpha}$ is a proper ideal in A. Let M be a maximal ideal in A containing $\sum R_{P_\alpha}$. Then $M \cap R$ contains a minimal prime ideal P_γ of R. Choose $a \in P_\gamma, a \notin \cup_{\beta\neq\gamma} P_\beta$. The element $a \in R$ is identified with $\{\theta_\alpha(a)\}$ in A. For $\beta \neq \gamma$, $\theta_\beta(a)$ is a unit in R_{P_β}. Define an element $u = \{u_\beta\} \in A$ by: $u_\beta = \theta_\beta(a)^{-1}$, if $\beta \neq \gamma$; and $u_\beta = 0$, if $\beta = \gamma$. Notice that $ua \in M$ is the identity at every component except the γ-component, where it is 0. Since $\sum R_{P_\alpha}$ is also contained in M, M must contain the identity of A. This contradiction completes the proof.

A Noetherian ring has the property that the annihilator of each ideal consisting entirely of zero divisors is nonzero [K, p. 56]. We abstract this to arbitrary commutative rings as follows: A ring R satisfies *Property A* if each finitely generated ideal $I \subseteq Z(R)$ has nonzero annihilator. For an element f in the polynomial ring $R[X]$, let $c(f)$ denote the *content* of f—that is, the ideal of R generated by the coefficients of f. Recall that if f is a zero divisor in a polynomial ring $R[X]$, then there is some nonzero element $a \in R$ such that $af = 0$, (McCoy's theorem).

Corollary 2.6 For a ring R the following conditions are equivalent:

(1) R has Property A.
(2) $T(R)$ has Property A.
(3) The element f is regular in $R[X]$ if and only if $c(f)$ is a regular ideal of R.

A ring T is *von Neumann regular* if for each a in T, there is some b in T such that $a^2b = a$. Clearly von Neumann regular rings are reduced. The following remark, an exercise in Gilmer's book [G, p. 111], will be useful in what follows.

Remark Let T be a reduced ring. Then the following conditions are equivalent:

(1) T is a von Neumann regular ring.
(2) T is a 0-dimensional ring.
(3) Each finitely generated ideal of T is principal and is generated by an idempotent.

PROOF: (1) $\Rightarrow$ (3): It suffices to prove that if $I = (a,b)$ is an ideal of T, then there exists an idempotent g in T such that $(g) = I$. Let $e = ax$ and $f = by$, where $a^2x = a$ and $b^2y = b$. Then $g = e + f - ef$ is the required element.

(3) $\Rightarrow$ (2): Assume $P_1 \subset P_2$ are prime ideals of T. Choose an idempotent $e \in P_2 \setminus P_1$. Then $e(1-e) = 0$, which implies that $1 - e \in P_1 \subset P_2$, a contradiction.

(2) $\Rightarrow$ (1): Let $a \in T$ and $I = \text{Ann}(a)$. Then $(I,a) \not\subseteq P$, for each prime ideal of T, Corollary 2.2. Thus $(I,a) = T$, so $1 = ax+c$ where $c \in I$. Therefore $a = a^2x$.

Von Neumann regular rings, being 0-dimensional, are total quotient rings. Note that any ring whose total quotient ring is von Neumann regular must have Property A.

A condition closely intertwined with Property A is the *annihilator condition*; abbreviated as (a.c.). A ring R satisfies (a.c.) if for each pair of elements a and b in R, there exists some c in R such that $\text{Ann}(a,b) = \text{Ann}(c)$. By induction, (a.c.) implies that if $a_1, \ldots, a_n \in R$, then there exists some c in R such that $\text{Ann}(a_1, \ldots, a_n) = \text{Ann}(c)$. The ring R satisfies (a.c.) if and only if its total quotient ring does. All von Neumann regular rings have (a.c.).

Property A and (a.c.) share many common features, but they are not equivalent. In fact, neither implies the other. Appropriate examples will be given later.

A *graded ring* means a ring graded by the integers. A graded ring is *nontrivial* in case it contains a regular homogeneous element of nonzero degree. The abundance of rings with Property A and (a.c.) can be seen by the next two results.

Theorem 2.7 Every nontrivial graded ring R satisfies Property A.

PROOF: Let x be a regular homogeneous element of nonzero degree in R and assume that $\deg x < 0$. The ring $R[1/x]$ can be made into a graded ring by defining $\deg(r/x^i) = \deg r - \deg x^i$, where r is a homogeneous element in R. The grade on $R[1/x]$ is compatible with the grade on R, and $R[1/x]$ contains a regular homogeneous element of positive degree. The presence of Property A in a ring R is determined by its total quotient ring. Therefore, assume that the graded ring R contains a regular homogeneous element x of degree $t > 0$.

Let $I = (a_1, \ldots, a_p)$ be an ideal contained in $Z(R)$. For $i = 1, \ldots, p$, let $a_i = \sum_{m_i}^{n_i} b_k^{(i)}$ be the homogeneous representation of a_i, where $\deg b_k^{(i)} = k$. Construct an element a as follows:

$$a = a_1 + a_2 x^{s_2} + \cdots + a_p x^{s_p}$$

where the s_i are integers such that $ts_2 + m_2 > n_1$ and $ts_i + m_i > n_{i-1} + ts_{i-1}$, for $i = 3, \ldots, p$.

Since $a \in Z(R)$, there exists a nonzero homogeneous element c such that $ca = 0$. (The proof of this is identical to the proof of McCoy's theorem—the statement immediately preceding Corollary 2.6.) By uniqueness of the representation of a by its homogeneous components $\{b_k^{(i)}x^{s_i}\}_{k,i}$, $cb_k^{(i)}x^{s_i} = 0$ for all i, k. Since x is not a zero divisor, $cb_k^{(i)} = 0$ for all i, k. This shows that $c \in \operatorname{Ann} I$ and that R satisfies Property A.

Theorem 2.8 If R is a reduced nontrivial graded ring, then R satisfies (a.c.).

PROOF: As in the proof of Theorem 2.7, nothing is lost by assuming that R contains a regular homogeneous element x of degree $t > 0$. Let (a, b) be an ideal in R. If $(a, b) \not\subseteq Z(R)$, then $\operatorname{Ann}(a, b) = \operatorname{Ann}(1)$. Assume that $(a, b) \subseteq Z(R)$ and write a and b in terms of their homogeneous components; say $a = a_m + \cdots + a_n$ and $b = b_h + \cdots + b_k$. Choose an integer s satisfying $h + st > n$ and let $c = a_m + \cdots + a_n + b_h x^s + \cdots + b_k x^s$.

Since R is a reduced ring, $\operatorname{Ann}(c) = \cap P$, where P varies over the minimal prime ideals of R not containing c. By [ZSII, p. 153], each P is a homogeneous ideal. Since the intersection of homogeneous ideals is homogeneous, $\operatorname{Ann}(c)$ is a homogeneous ideal.

Let d be a homogeneous element in $\operatorname{Ann}(c)$. Then $da_i = 0$ and $db_j x^s = 0$ for all i, j. Thus $da = 0 = db$ which implies that $\operatorname{Ann}(c) \subseteq \operatorname{Ann}(a, b)$. The other inclusion is obvious.

Corollary 2.9 The polynomial ring $R[X]$ satisfies Property A, and if R is reduced $R[X]$ satisfies (a.c.).

Material concerning the complete ring of quotients of a commutative ring R may be found in Lambek's book [L, pp. 36–46]. For completeness we give a summary of the pertinent definitions and facts about such rings. A subset S of R is *dense* if $\operatorname{Ann} S = 0$. If D_1 and D_2 are two dense ideals and if $f_i \in \operatorname{Hom}_R(D_i, R)$, then

$D_1 \cap D_2$ and $f_2^{-1} D_1 = \{r \in R: f_2 r \in D_1\}$ are also dense. Define

$$(-f_1)d = -(f_1 d)$$
$$f_1 + f_2 \in \mathrm{Hom}_R(D_1 \cap D_2, R) \text{ by } (f_1 + f_2)d = f_1 d + f_2 d$$
$$f_1 f_2 \in \mathrm{Hom}_R(f_2^{-1} D_1, R) \text{ by } (f_1 f_2)d = f_1(f_2 d)$$
$$0, 1 \in \mathrm{Hom}_R(R, R) \text{ by } 0r = 0 \text{ and } 1r = r$$

The elements $f_i \in \mathrm{Hom}_R(D_i, R)$ are called *fractions.* An equivalence relation θ is defined on the set of fractions by $f_1 \theta f_2$ if and only if f_1 and f_2 agree on $D_1 \cap D_2$. The set of all equivalence classes forms a commutative ring $Q(R)$ called the *complete ring of quotients* of R. An element $a/b \in T(R)$ may be considered as an element of $\mathrm{Hom}_R(bR, R)$ by defining $(a/b) : bR \to R$ by $a/b(br) = ar$. The mapping $a/b \mapsto \theta(a/b)$ is an embedding of $T(R)$ into $Q(R)$. Thus, without loss of generality, assume that $R \subseteq T(R) \subseteq Q(R)$. The proof of the next theorem may be found in [L, p. 42].

Theorem 2.10 Let R be a ring. Then $Q(R)$ is a von Neumann regular ring if and only if R is reduced.

When studying Property A it is sufficient to limit the discussion to total quotient rings, Corollary 2.6. The reduced total quotient rings with Property A are characterized below.

Theorem 2.11 For a reduced total quotient ring T, the following statements are equivalent:

(1) T has Property A.
(2) If I is a finitely generated proper ideal of T, then I is contained in a minimal prime ideal of T.
(3) If I is a finitely generate proper ideal of T, then I extends to a proper ideal in $Q(T)$.

PROOF: (1) $\Leftrightarrow$ (2): Apply Corollary 2.3.

(1) $\Rightarrow$ (3): The annihilator of an ideal I of T also annihilates $IQ(T)$.

(3) $\Rightarrow$ (1): Assume that I is a finitely generated dense ideal of T such that $I \subseteq Z(T)$. Then I is a dense subset of $Q(T)$ [L, p. 41]; whence $IQ(T)$ is a dense ideal of $Q(T)$. But $Q(T)$ is a von Neumann regular ring, Theorem 2.10. So $Q(T)$ has Property A. By the equivalence of (1) and (2), $IQ(T)$ is not contained in any minimal prime ideal of $Q(T)$. However, in von Neumann regular rings, minimal primes are maximal. Therefore $IQ(T) = Q(T)$, a contradiction.

Corollary 2.12 Every zero-dimensional ring satisfies Property A.

PROOF: Let I be a finitely generated proper ideal of a zero-dimensional ring T. Then $I \subseteq P$, where P is a minimal prime ideal of T. If $I \subseteq N(T)$, then $\operatorname{Ann} I \supseteq I^{n-1} \neq (0)$ for some $n \geq 2$. If $I \not\subseteq N(T)$, then $\pi(I) \subseteq \pi(P)$ where $\pi : T \to T/N(T)$ is the canonical homomorphism. There exists some $z \subseteq T \setminus N(T)$ such that $\pi(z)\pi(I) = \pi(0)$; that is, $zI \subseteq N(T)$. Therefore $\operatorname{Ann} I \supseteq z^n I^{n-1} \neq (0)$ for some $n \geq 2$.

3. Zero-Dimensional Rings

A reduced ring R can always be embedded into a von Neumann regular ring, namely, $Q(R)$. Von Neumann regular rings are zero-dimensional and reduced; hence a ring with nontrivial nilpotent elements cannot possibly be embedded into a von Neumann regular ring. What is the proper generalization of the reduced case to a ring with a nonzero nilradical? This section gives one answer to this question by characterizing those rings that may be embedded into 0-dimensional rings. Along the way we give characterizations of von Neumann regular rings and 0-dimensional rings.

A ring R is *π-regular* if for each $r \in R$ there is a positive integer n and an element $x \in R$ such that $r^{2n}x = r^n$. The set of π-regular rings in the class of nonreduced rings is analogous to the set of von Neumann regular rings in the class of reduced rings.

Theorem 3.1 Let R be a ring. Then the following conditions are equivalent:

(1) R is π-regular.

(2) For each $r \in R$, there is an element $y \in R$ and a positive integer n such that $r^{n+1}y = r^n$.

(3) $R/N(R)$ is a von Neumann regular ring.

(4) R is 0-dimensional.

PROOF: (1) $\Rightarrow$ (2): Let $y = r^{n-1}x$.

(2) $\Rightarrow$ (1): Since $r^n = r^{n+1}y = r^n(ry) = r^{n+1}yry = r^{n+2}y^2 = \cdots = r^{2n}y^n$, choose $x = y^n$.

(2) $\Rightarrow$ (3): Let r be in R and let $\bar{r}$ be its residue in $R/N(R)$. Then $\bar{r}^{n+1}\bar{y} = \bar{r}^n$ is in $R/N(R)$ for appropriate y and n. It follows that $(\bar{r}^n\bar{y} - \bar{r}^{n-1})^2 = \bar{0}$, and since $R/N(R)$ is a reduced ring, $\bar{r}^n\bar{y} = \bar{r}^{n-1}$. Continuing this process yields $\bar{r}^2\bar{y} = \bar{r}$.

(3) $\Rightarrow$ (2): For $r \in R$ choose $x \in R$ such that $r - r^2x \in N(R)$. Then $0 = (r - r^2x)^n = r^n(1 - rx)^n$ for some positive integer n. Write $(1 - rx)^n = 1 - prx$ and define $y = px$.

(3) $\Leftrightarrow$ (4): The rings R and $R/N(R)$ are simultaneously 0-dimensional.

Zero-dimensional rings are total quotient rings. Thus the next result may be stated only for such rings.

Theorem 3.2 Let T be a total quotient ring. Then the following conditions are equivalent:

(1) T is π-regular.

(2) For each $a \in T$, there exists an idempotent $e_a \in T$ such that $a + (1 - e_a)$ is a unit and $a(1 - e_a)$ is nilpotent.

(3) For each $a \in T$, there exists $b \in T$ such that $a + b$ is a unit and ab is nilpotent.

(4) For each $a \in T$, there exists a positive integer n such that $a^n = re$, where r is a unit and e is idempotent.

PROOF: (1) ⇒ (2): Let a be in T. Choose some a' in T such that $a^{2n}a' = a^n$. It is easy to verify that $a^n a'$ and $1 - a^n a'$ are idempotent elements in T. Furthermore $[a(1 - a^n a')]^n = a^n(1 - a^n a')^n = a^n(1 - a^n a') = 0$; hence $a(1 - a^n a')$ is nilpotent. To show that $a + (1 - a^n a')$ is a unit element, it suffices to show that it does not belong to any prime ideal P of T. If P is a prime ideal of T, then $a(1 - a^n a') \in P$. Both a and $1 - a^n a'$ cannot be in P; for if so $1 \in P$. Thus $a + (1 - a^n a') \notin P$.

(2) ⇒ (3): This implication is clear.

(3) ⇒ (4): Let a be in T and choose b satisfying the requirements of (3). If P is a prime ideal of T and if $a^n + b^n \in P$ where n is a positive integer such that $(ab)^n = 0$, then a^n and b^n belong to P. Therefore, a and b are in P, contradicting the hypothesis that $a + b$ is a regular element. If $r = a^n + b^n$ and $e = a^n/(a^n + b^n)$, then r and e are the required elements.

(4) ⇒ (1): Let $P_1 \subseteq P_2$ be prime ideals of T. If $a \in P_2$, then $a^n = re$; where n, r, and e are prescribed by (4). The element $1 - e \notin P_2$ and $e(1 - e) = 0$ implies that $e \in P_1$. Therefore $P_1 = P_2$, which shows that T is 0-dimensional.

Specializing Theorem 3.2 to reduced total quotient rings yields characterizations of von Neumann regular rings.

Corollary 3.3 Let T be a reduced total quotient ring. Then the following conditions are equivalent:

(1) T is von Neumann regular.

(2) For each $a \in T$, there exists an idempotent $e_a \in T$ such that $a + (1 - e_a)$ is a unit and $a(1 - e_a) = 0$.

(3) For each $a \in T$, there exists $b \in T$ such that $a + b$ is a unit and $ab = 0$.

(4) For each $a \in T$, $a = re$ where r is a unit and e is idempotent.

There are many other characterizations of π-regular rings and von Neumann regular rings. The characterizations given here will suffice for this book.

Let Spec R be the set of all prime ideals of R and denote the maximal elements of Spec R by Max R.

Lemma 3.4 Let T be a 0-dimensional ring. For each $M_\lambda \in \operatorname{Max} T$, let $S_{M_\lambda}(0)$ be the kernel of the canonical homomorphism $T \to T_{M_\lambda}$. Then $S_{M_\lambda}(0)$ is M_λ-primary and $\cap S_{M_\lambda}(0) = (0)$.

PROOF: Since $S_{M_\lambda}(0) = 0_{M_\lambda} \cap T$, it is M_λ-primary.

For the second part, let $x \in \cap S_{M_\lambda}(0)$. Then for each λ there exists $s_\lambda \in T \setminus M_\lambda$ such that $xs_\lambda = 0$. This proves that the $\operatorname{Ann}(x)$ is not contained in any maximal ideal of T. Therefore $x = 0$.

Theorem 3.5 Let R be a commutative ring. Then there exists a 0-dimensional ring $\tilde{R}$ containing R as a subring if and only if there exists a family $\{Q_\lambda\}$ of primary ideals of R satisfying the following conditions: (1) $\cap Q_\lambda = (0)$; and (2) if $a \in R$ then there exists a positive integer n_a such that $a^{n_a} \notin \cup(P_\lambda \setminus Q_\lambda)$, where P_λ is the prime ideal associated with Q_λ.

PROOF: ($\Rightarrow$): Assume that R can be embedded into a 0-dimensional ring $\tilde{R}$. If $a \in R$, choose an idempotent element $e \in \tilde{R}$ so that $a+(1-e)$ is a unit and $a(1-e)$ is nilpotent, Theorem 3.2. Let $\{M_\alpha\}$ be the set of prime ideals of $\tilde{R}$ containing a. Then $1 - e \notin M_\alpha$, for each α. Let S be the multiplicatively closed set $\tilde{R} \setminus (\cup M_\alpha)$, and consider the ring $\tilde{R}_S$. If $r \in \tilde{R}$, write $r/1$ for the image of r in $\tilde{R}_S$. It is clear that $(1 - e)/1$ is a unit in $\tilde{R}_S$. It follows that there exists a positive integer n_a such that $a^{n_a} \in \cap S_{M_\alpha}(0)$, where

$S_{M_\alpha}(0)$ is the kernel of the canonical homomorphism $\tilde{R} \to \tilde{R}_{M_\alpha}$. For each $M_\lambda \in \operatorname{Spec} \tilde{R}, Q_\lambda = S_{M_\lambda}(0) \cap R$ is $(M_\lambda \cap R)$-primary and $\cap Q_\lambda = (0)$. Furthermore $a^{n_a} \notin M_\lambda \cap R \setminus Q_\lambda$, for each λ.

($\Leftarrow$): Let $\{Q_\lambda\}$ be the set of P_λ-primary ideals satisfying conditions (1) and (2). Let $\bar{R}$ be the complete direct product of the family $\{(R/Q_\lambda)_{P_\lambda/Q_\lambda}\}$. Define a map $\varphi : R \to \bar{R}$ by

$$\varphi(r) = \left(\frac{r + Q_\lambda}{1 + Q_\lambda}\right)_\lambda$$

Condition (1) shows that φ is a monomorphism. Hence R may be assumed to be a subring of $\bar{R}$. For $a \in R$, define e_a to be the idempotent of $\bar{R}$ such that the λth component of e_a is

$$\begin{cases} 1 & \text{if } a \notin P_\lambda \\ 0 & \text{if } a \in P_\lambda \end{cases}$$

It is easy to see that $a + (1 - e_a)$ is a regular element of $\bar{R}$, and condition (2) implies that $a(1 - e_a)$ is nilpotent. Let R_1 be the subring of $\bar{R}$ generated by R and $\{e_a : a \in R\}$, and define $\tilde{R}$ to be the total quotient ring or R_1.

Assume that $P_1 \subset P_2$ are prime ideals of $\tilde{R}$. Then $P_1 \cap R_1 \subset P_2 \cap R_1$. Certainly R_1 is integral over R (for each $e_a, e_a^2 - e_a = 0$), so $P_1 \cap R \subset P_2 \cap R$. Let $r \in P_2 \cap R, r \notin P_1 \cap R$, and let e_r be the idempotent element of $\tilde{R}$ such that $r(1 - e_r)$ is nilpotent and $r + (1 - e_r)$ is regular. It follows that $1 - e_r \in P_1$, and hence $r + (1 - e_r) \in P_2$, a contradiction. This completes the proof.

Corollary 3.6 If R is a ring in which (0) has a (finite) primary decomposition, then R can be embedded into a 0-dimensional ring. In particular, each Noetherian ring can be embedded into a 0-dimensional ring.

It is instructive to use the proof of Theorem 3.5 to compute $\tilde{Z}$ for the ring of integers Z, where the family of primary ideals is taken to be the set of squares of proper prime ideals of Z. Let $\{P_\lambda^2\}$ be this set and let $\bar{Z} = \prod (Z/P_\lambda^2)_{P_\lambda/P_\lambda^2} = \prod Z/P_\lambda^2$. Let $\varphi : Z \to \bar{Z}$ as in Theorem 3.5. Then

$$\varphi(3) = (3 + (2^2), 3 + (3^2), 3 + (5^2), 3 + (7^2), \ldots)$$

Note that $\varphi(3)$ is a zero divisor of $\bar{Z}$, since $0 = \varphi(3)(0 + (2^2), 3 + (3^2), 0 + (5^2), \ldots)$. Therefore $\varphi(3)$ has no inverse in $\tilde{Z}$ (since it has none in $\bar{Z}$). If Q is the usual quotient field of Z, then $\tilde{Z}$ contains no isomorphic copy of Q.

We end this section with an example of a ring that cannot be embedded in a 0-dimensional ring. Let V be a 2-dimensional valuation domain. Let $M \supset P$ be the nonzero prime ideals of V, where M is maximal. Choose a nonzero element a in P, and set $A = aM$. Define $R = V/A$. The element $\bar{a} = a + A \in R$ is nonzero and is contained in every nonzero ideal of R. Furthermore, the zero ideal of R is not primary; for if $x \in M \setminus P$, then $\bar{x}\bar{a} = \bar{0}$ where $\bar{x} = x + A$. This proves that no collection of primary ideals of R satisfies condition (1) of Theorem 3.5; that is, R is not embeddable in a 0-dimensional ring.

4. Compactness of Min R

Let R be a commutative ring. If $a \in R$, let $X_a = \{P \in \operatorname{Spec} R : a \notin P\}$. A topology on $\operatorname{Spec} R$ may be defined by taking as the basic open sets $\{X_a : a \in R\}$; see [L, pp. 47–48]. This is called the *Zariski topology* on $\operatorname{Spec} R$. The set $\operatorname{Min} R$ may be treated as a subspace of $\operatorname{Spec} R$. While $\operatorname{Spec} R$ is always a compact topological space, $\operatorname{Min} R$ may or may not be compact. This section completely determines the compactness of $\operatorname{Min} R$. If H is a subset of R, define $S(H) = \{M \in \operatorname{Spec} Q(R) : H \not\subseteq M \cap R\}$.

The next theorem shows that when discussing the compactness of Min R it is enough to consider reduced total quotient rings.

Theorem 4.1 Let R be a commutative ring. The following conditions are equivalent:

(1) $\operatorname{Min} R$ is compact.
(2) $\operatorname{Min} R/N(R)$ is compact.
(3) $\operatorname{Min} T(R/N(R))$ is compact.

PROOF: Write N for $N(R)$. The map $P \mapsto P/N$ is a homeomorphism from $\operatorname{Min} R$ onto $\operatorname{Min} R/N$, and the map $P/N \mapsto (P/N)T(R/N)$ is a homeomorphism between $\operatorname{Min} R/N$ and $\operatorname{Min} T(R/N)$.

The basic open sets of the topological space $\operatorname{Min} R$ will be denoted by X_a, while those of $\operatorname{Spec} Q(R)$ will be denoted by X'_a. As usual, $Q(R)$ is the complete ring of quotients of R.

Lemma 4.2 Let R be a reduced ring. Then

$$\operatorname{Min} R \subseteq \{M \cap R : M \in \operatorname{Spec} Q(R)\}$$

PROOF: Let P be a minimal prime ideal of R. Assume that $M \cap R \not\subseteq P$, for each $M \in \operatorname{Spec} Q(R)$. Let $t \in R \setminus P$ and define $\mathcal{U}_t = \{M \in \operatorname{Spec} Q(R) : t \in (M \cap R) \setminus P\}$. There is an idempotent e in the von Neumann regular ring $Q(R)$ such that e and t generate the same principal ideal of $Q(R)$. It follows that the basic open set X'_{e-1} of $\operatorname{Spec} Q(R)$ is the same as $\mathcal{U}_t$; that is, $\mathcal{U}_t$ is open. Since $\{\mathcal{U}_t : t \in R \setminus P\}$ is an open cover of the compact space $\operatorname{Spec} Q(R)$, there exist $t_1, \ldots, t_n \in R \setminus P$ such that $\{\mathcal{U}_{t_1}, \ldots, \mathcal{U}_{t_n}\}$ is a finite subcover of $\operatorname{Spec} Q(R)$. This leads to the contradictory statements $t = t_1 t_2 \cdots t_n \in R \setminus P$ and $t \in \cap\{M : M \in \operatorname{Spec} Q(R)\} = (0)$. Therefore there exists some $M \in \operatorname{Spec} Q(R)$ such that $M \cap R = P$ and hence $\operatorname{Min} R \subseteq \{M \cap R : M \in \operatorname{Spec} Q(R)\}$.

Theorem 4.3 For a reduced ring R, the following statements are equivalent:

(1) $\operatorname{Min} R$ is compact.
(2) $\{M \cap R : M \in \operatorname{Spec} Q(R)\} = \operatorname{Min} R$.
(3) If $a \in R$, then there exists a finitely generated ideal I of R such that $\operatorname{Spec} Q(R) \setminus S(a) = S(I)$.
(4) If $b \in R$, then there exists a finitely generated ideal $J \subseteq \operatorname{Ann}(b)$ such that $\operatorname{Ann}(J, b) = (0)$.
(5) $Q(R)$ is a flat R-module.

PROOF: (2) $\Rightarrow$ (1): $\operatorname{Min} R$ is a continuous image of $\operatorname{Spec} Q(R)$ via the map $M \mapsto M \cap R$. Since $\operatorname{Spec} Q(R)$ is compact, $\operatorname{Min} R$ is compact.

(3) $\Rightarrow$ (2): In view of Lemma 4.2, it suffices to prove that if $M \in \operatorname{Spec} Q(R)$, then $M \cap R$ is a minimal prime ideal of R. Suppose not; say that $M \cap R$ is not a minimal prime for some $M \in \operatorname{Spec} Q(R)$. Then there is some prime P in $\operatorname{Spec} Q(R)$ such that $P \cap R \subset M \cap R$. If $a \in M \cap R \setminus P \cap R$, then $P \in S(a)$ and $M \notin S(a)$. Let I be the finitely generated ideal of R such that $\operatorname{Spec} Q(R) \setminus S(a) = S(I)$. Then $M \in S(I)$ and $P \notin S(I)$, which implies that $I \nsubseteq M \cap R$, but $I \subseteq P \cap R \subseteq M \cap R$, a contradiction.

(1) $\Rightarrow$ (4): If $b \in R$, then $(\operatorname{Ann}(b), b)$ is not contained in any minimal prime ideal of R, Corollary 2.2. Thus $\operatorname{Min} R = X_b \cup (\cup\{X_a : a \in \operatorname{Ann}(b)\})$. Since $\operatorname{Min} R$ is compact, there exist $a_1, \ldots, a_n \in \operatorname{Ann}(b)$ such that $\operatorname{Min} R = X_b \cup X_{a_1} \cup \cdots \cup X_{a_n}$. Let $J = (a_1, \ldots, a_n)$, then $\operatorname{Ann}(J, b) = (0)$.

(4) $\Rightarrow$ (3): It suffices to prove that if J is the finitely generated ideal of the hypothesis, then $\operatorname{Spec} Q(R) \setminus S(b) = S(J)$. If $M \in \operatorname{Spec} Q(R) \setminus S(b)$, then $b \in M \cap R$. But (J, b) is dense in R implies (J, b) is dense in $Q(R)$ [L, p. 41]; hence $J \nsubseteq M \cap R$. Thus $\operatorname{Spec} Q(R) \setminus S(b) \subseteq S(J)$. On the other hand, take $M \in S(J)$. Then $J \nsubseteq M \cap R$. Since $bJ = 0$, $b \in M \cap R$; or equivalently, $M \in \operatorname{Spec} Q(R) \setminus S(b)$.

(5) $\Rightarrow$ (3): Let $a \in R$. By flatness, the map $Q(R) \otimes_R (a) \rightarrow$

$Q(R)$ is injective. Choose $e \in Q(R)$ such that $ea = 0$ and $e + a$ is a unit of $Q(R)$, Corollary 3.3. Consequently $\operatorname{Spec} Q(R) \setminus S(a) = \mathcal{U}$, where $\mathcal{U}$ is the set of maximal ideals of $Q(R)$ not containing e. Since $e \otimes a = 0$, there exist families $\{q_i\}_{i=1}^n \subseteq Q(R)$ and $\{r_i\}_{i=1}^n \subseteq R$ such that $r_j a = 0$ and $e = \sum q_i r_i$ [B, p. 25]. Obviously $S(r_i) \subseteq \operatorname{Spec} Q(R) \setminus S(a)$. If $I = (r_1, \ldots, r_n)$, then $\mathcal{U} \subseteq \cup S(r_i) = S(I) \subseteq \operatorname{Spec} Q(R) \setminus S(a) = \mathcal{U}$. Therefore $S(I) = \operatorname{Spec} Q(R) \setminus S(a)$.

(3) $\Rightarrow$ (5): It is enough to prove that $Q(R) \otimes_R I \to IQ(R)$ is an injective map for each finitely generated ideal I of R. Suppose not; assume that $I = (a_1, \ldots, a_n)$ is a minimally generated ideal with $\{a_i\}_{i=1}^n$ a set of minimal generators for I such that $Q(R) \otimes_R I \to IQ(R)$ is not one-to-one. Let $x = \sum_{j=1}^n q_j \otimes a_j \neq 0$ with $\sum_{j=1}^n q_j a_j = 0$.

Consider one of the generators a_i of I. Then $a_i x = a_i \sum_{j=1}^n q_j \otimes a_j = \sum_{j=1}^n q_j \otimes a_i a_j = \sum_{j=1}^n q_j a_j \otimes a_i = 0 \otimes a_i = 0$. Therefore $I \subseteq \operatorname{Ann}_{Q(R)}(x) = \cap M_\beta$, where $M_\beta \in \operatorname{Spec} Q(R)$ and $x \notin M_\beta$. By the hypothesis, there exists a finitely generated ideal J of R such that $\operatorname{Spec} Q(R) \setminus S(a_n) = S(J)$. If the ideal J is contained in $\operatorname{Ann}_{Q(R)}(x)$, then J and a_n belong to a common $M \in \operatorname{Spec} Q(R)$; that is, $M \notin S(J)$ and $M \notin S(a_n)$. This is a contradiction. Hence we may choose some $b \in J$ such that $bx \neq 0$ and $S(b) \cap S(a_n) = \emptyset$. The last equality shows that $ba_n = 0$. Let $I' = (a_1, \ldots, a_{n-1})$ and note that $Q(R) \otimes_R I' \to IQ(R)$ is not injective since $\sum_{j=1}^{n-1} q_j b \otimes a_j \neq 0$, yet $\sum_{j=1}^{n-1} q_j b a_j = b(\sum_{j=1}^n q_j a_j) = 0$. This contradicts the assumption that I is minimally generated. Therefore $Q(R)$ is an R-flat module.

The proof of Theorem 4.3 is now complete.

Because of the techniques used, a direct proof of (1) $\Rightarrow$ (2) of the above theorem is instructive. We give it here.

PROOF: (1) $\Rightarrow$ (2): Assume that (2) is not true. Then there exists $M \in \operatorname{Spec} Q(R)$ such that $M \cap R$ is not a minimal prime ideal. Define $\mathcal{U} = \{M_0 \in \operatorname{Spec} Q(R) \colon M_0 \cap R \in \operatorname{Min} R\}$. If

$P \in \operatorname{Min} R$, choose $b \in M \cap R \setminus P$. Therefore $P \in X_b$. This proves that $\operatorname{Min} R = \cup X_b$, where b varies over the set $M \cap R$. By compactness $\operatorname{Min} R = \cup_{i=1}^{n} X_{b_i}$, for some $b_1, \ldots, b_n \in M \cap R$. If $I = (b_1, \ldots, b_n)$, then $S(I) \supseteq \mathcal{U}$. The ideal $IQ(R)$ is finitely generated by an idempotent element e. Thus $S(I)$ is equal to the basic open set X'_e of $\operatorname{Spec} Q(R)$. Also $\operatorname{Spec} Q(R) \setminus S(I) = X'_{1-e}$, so $S(I)$ is both open and closed in $Q(R)$. Note that $e \neq 1$, for otherwise $\operatorname{Spec} Q(R) = S(I)$, but $M \in \operatorname{Spec} Q(R) \setminus S(I)$. Define $J = \{q \in Q(R): X'_q \subseteq \operatorname{Spec} Q(R) \setminus S(I)\}$. Then J is an ideal; and since $1 - e \in J$, $J \neq (0)$.

Suppose, for the moment, that $J \cap R = (0)$. If $q \in J$, then $D = q^{-1}R = \{r \in R: qr \in R\}$ is a dense subset in $Q(R)$ [L, p. 40]. Hence $qD \subseteq R \subseteq J \cap R = 0$, which implies $q = 0$. Therefore $J = (0)$. This contradiction proves that $J \cap R \neq (0)$. Let c be a nonzero element in $J \cap R$. Then $S(c) \subseteq \operatorname{Spec} Q(R) \setminus S(I)$, and hence $c \in M_0 \cap R$, for all $M_0 \in S(I) \supseteq \mathcal{U}$. By Lemma 4.2, $\{M_0 \cap R: M_0 \in \mathcal{U}\} \supseteq \operatorname{Min} R$. Since R is a reduced ring, $c \in \cap\{M_0 \cap R: M_0 \in \mathcal{U}\} = (0)$, a contradiction. Therefore $M \cap R$ is a minimal prime ideal of R, for each $M \in \operatorname{Spec} Q(R)$. Using this, along with Lemma 4.2, establishes (2).

A ring R is *rationally complete* if $R \cong Q(R)$.

Corollary 4.4 If R is a rationally complete reduced ring, then $\operatorname{Min} R$ is compact.

The next result connects the concept of $\operatorname{Min} R$ being compact with that of $T(R)$ being a von Neumann regular ring.

Theorem 4.5 Let R be a reduced ring with total quotient ring $T(R) = T$. The following conditions are equivalent:

(1) T is a von Neumann regular ring.
(2) R satisfies Property A and $\operatorname{Min} R$ is compact.

(3) R satisfies (a.c.) and $\operatorname{Min} R$ is compact.

(4) Each finitely generated ideal of R contained in a union of minimal prime ideals of R is contained in one of them, and $\operatorname{Min} R$ is compact.

PROOF: Since Property A, (a.c.), $\operatorname{Min} R$ is compact, and the first part of (4) hold in R if and only if they hold in $T(R)$, it is sufficient to prove the theorem under the assumption that $R = T(R) = T$.

(1) $\Rightarrow$ (2): Use Corollary 2.12 and [L, p. 47].

(2) $\Rightarrow$ (1): It suffices to prove that $\dim T = 0$. Assume that $Q_1 \supset Q_0$ are prime ideals of T with Q_0 minimal. Choose $a \in Q_1 \setminus Q_0$. By Theorem 4.3, there exists a finitely generated ideal I of T such that $\operatorname{Spec} Q(T) \setminus S(a) = S(I)$. Let $P \in \operatorname{Min} T$ and assume that $(I, a) \subseteq P$. For some $M \in \operatorname{Spec} Q(T)$, $M \cap T = P$, Lemma 4.2. But this last statement implies that $M \notin S(a)$ and $M \notin S(I)$, a contradiction. Thus (I, a) is contained in no minimal prime ideal of T. Therefore $\operatorname{Ann}(I, a) = (0)$, Corollary 2.3. On the other hand there is some $M_0 \in \operatorname{Spec} Q(T)$ such that $M_0 \cap T = Q_0$; so $I \subseteq Q_0$, and hence $(I, a) \subseteq Q_1$. By Property A, $\operatorname{Ann}(I, a) \neq 0$. The only way out of this dilemma is for T to be 0-dimensional.

(1) $\Rightarrow$ (3): This direction is clear.

(3) $\Rightarrow$ (1): Again it is sufficient to prove that $\dim T = 0$. Suppose that $M \supset Q$ are prime ideals of T with Q minimal. If $a \in M \setminus Q$, then there exists a finitely generated ideal I of T such that $\operatorname{Spec} Q(T) \setminus S(a) = S(I)$. Let $b \in T$ such that $\operatorname{Ann}(b) = \operatorname{Ann} I$. Then $S(b) = S(I)$. For each $P \in \operatorname{Min} T$, exactly one of the following elements a or b is in P. In particular, $b \in Q$. Thus $a + b \in M \subseteq Z(T)$. This contradicts the fact that $a + b \notin P$ for each $P \in \operatorname{Min} T$; for $a + b$ is a zero divisor, so it must lie in some P, Corollary 2.4.

(1) $\Rightarrow$ (4): This is clear.

(4) $\Rightarrow$ (2): The ring T has Property A by Theorem 2.11.

A nice characterization of the compactness of $\operatorname{Min} R$ may be derived from Theorem 4.5.

Corollary 4.6 Let R be a reduced ring. Then $\operatorname{Min} R$ is compact if and only if $T(R[X])$ is a von Neumann regular ring.

PROOF: The map $\operatorname{Min} R$ onto $\operatorname{Min} R[X]$ given by $P \mapsto P[X]$ is a homeomorphism.

($\Rightarrow$): $\operatorname{Min} R[X]$ is compact, hence $\operatorname{Min} T(R[X])$ is compact. By Corollary 2.9, $T(R[X])$ has Property A. Apply Theorem 4.5.

($\Leftarrow$): Clear.

Let R be a ring. Then R is *coherent* if for each finitely generated ideal I of R, there exists an exact sequence $R^m \to R^n \to I \to 0$. A coherent ring has the property that for each $x \in R$, $\operatorname{Ann}(x)$ is a finitely generated ideal of R [B, pp. 44–5].

Theorem 4.7 For a reduced coherent ring R, the following conditions are equivalent:

(1) R has Property A.

(2) R has (a.c.).

(3) $T(R)$ is a von Neumann regular ring.

PROOF: (1) $\Rightarrow$ (3): In view of Theorem 4.5, it is sufficient to prove that $\operatorname{Min} R$ is compact. Let $x \in R$; then $\operatorname{Ann}(x) = I$ is a finitely generated ideal of R. Assume $I \subseteq M$ for some $M \in \operatorname{Spec} Q(R) \setminus S(x)$. Then $(I, x) \subseteq M \cap R \subseteq Z(R)$. By Property A $\operatorname{Ann}(I, x) \neq 0$. But this contradicts the fact that $(I, x) \not\subseteq P$, for each $P \in \operatorname{Min} R$. It follows that $\operatorname{Spec} Q(R) \setminus S(x) = S(I)$. By Theorem 4.3, $\operatorname{Min} R$ is compact.

(2) $\Rightarrow$ (3): For any $x \in R$, $\operatorname{Ann}(x) = (z_1, \ldots, z_n)$, and $\operatorname{Ann}(\operatorname{Ann}(x)) = \operatorname{Ann}(z)$ for appropriate $z_1, \ldots, z_n, z \in R$. Thus $\operatorname{Ann}(x) = \operatorname{Ann}(\operatorname{Ann}(z))$.

Let Q be a prime ideal of R consisting entirely of zero divisors. We show that $Q \in \operatorname{Min} R$. Assume that $x \in Q$. If $P \in \operatorname{Min} R$, then

$x \in P$ if and only if $z \notin P$, where z is constructed as above. To see this suppose that x and z are both in P, then $\mathrm{Ann}(\mathrm{Ann}(z)) \subseteq P$; but $\mathrm{Ann}(x) \not\subseteq P$, a contradiction. A similar contradiction arises if neither x nor z is assumed to be in P. Thus $x + z$ is a regular element in R, Corollary 2.4. Therefore $x + z \notin Q$, so $z \notin Q$. But $zx = 0$. By Corollary 2.2, $Q \in \mathrm{Min}\, R$.

Since there is a one-to-one order preserving correspondence between the prime ideals of R contained in $Z(R)$ and the prime ideals of $T(R), T(R)$ is 0-dimensional; hence a von Neumann regular ring.

(3) $\Rightarrow$ (1), (2): Clear.

Notes

Section 2 Versions of (2.1) appear in several places, including the Exercises in [B]. Our statement of (2.1) is from Kist [60]. Matlis proved (2.5) [78]. Theorems (2.7), (2.8), and (2.11) come from Huckaba and Keller [55]. The statement and proof of McCoy's Theorem, referred to in Section 2, may be found in [70, p. 34].

Section 3 Some of the characterizations of von Neumann regular rings given in this section, as well as the Remark in Section 2, were implicit in Endo's work [29]. Theorem 3.1 is a result of Storrer [103]. Most of (3.2) and (3.3) on von Neumann regular and π-regular rings are due to Arapovic [10]. Many more characterizations of von Neumann regular and π-regular rings are known. The best source of information on von Neumann regular rings—mostly noncommutative, but some commutative—is Goodearl's book [40]. The embedding result, (3.5) is by Arapovic [12].

There are many other aspects of zero dimensional rings that have not been included in Section 3. Noteable are Pierce's work on von Neumann regular rings which appeared in his Memoirs [92];

and the papers by R. Wiegand [105], Fontana and Mazzola [32] and [33], Maroscia [74], and Olivier [90].

Section 4 Mewborn was the first to characterize when $\operatorname{Min} R$ is compact. He proved the equivalence of (1), (2), (3), and (5) of (4.3) [85]. Conditions (1) and (4) of (4.3) are shown to be equivalent by Quentel [96]. In addition this paper also gives the equivalence of (1) and (2) of (4.5), and the characterization of compactness of $\operatorname{Min} R$ that is given in (4.6). An important early paper on this subject is that of Henricksen and Jerison [49]. This paper initiated the study of the compactness of $\operatorname{Min} R$; and, in fact, proved the equivalence of (1) and (3) of (4.5), as well as (2) implies (3) of (4.3). In a more recent paper [91] Picavet gives a topological proof of parts of (4.3). Theorem 4.7 is from [55]. There is more about $\operatorname{Min} R$ to be said than has been presented here. Hochster [52] defines a topological space to be minspectral if it is homeomorphic to $\operatorname{Min} R$ for some ring R (possibly without identity). He characterizes these spaces, shows that each can arise from a ring with identity, and shows how to construct such spaces. Additional information, properties, applications, and related topics about $\operatorname{Min} R$ may be found in [78], [91], and the paper by Popescu and Vraciu [93].

Chapter II

Valuation Theory

5. Valuations

In the late 1950s P. Samuel gave several possible generalizations of valuation theory to arbitrary commutative rings. During the next twenty years other generalizations followed. Since these new extensions are not equivalent, a choice must be made as to the best way to generalize the classical theory. The following points of view are taken here: (1) The valuation maps should be defined exactly as in the classical case; and (2) The valuation rings should arise from the valuation maps as in the integral domain case.

With these restrictions in mind, this section develops the theory of valuations for rings with zero divisors.

Let $(G, +)$ be a totally ordered abelian group. Extend G by the symbol ∞ by defining $g < \infty$, $g + \infty = \infty + \infty = \infty$ for all

$g \in G$. Leave $\infty - \infty$ undefined. A *valuation* on a ring T with *value group* G is a mapping v from T *onto* $G \cup \{\infty\}$ such that:

(a) $v(xy) = v(x) + v(y)$, for all x and y in T.
(b) $v(x + y) \geq \min\{v(x), v(y)\}$, for all x and y in T.
(c) $v(1) = 0$ and $v(0) = \infty$.

Let v be a valuation on T. It is easily seen that: (i) if x is a unit in T, then $v(1/x) = -v(x)$; and (ii) if $v(x) \neq v(y)$, then $v(x + y) = \min\{v(x), v(y)\}$.

Theorem 5.1 Let R be a subring of the ring T, and let P be a prime ideal of R. Then the following conditions are equivalent:

(1) If B is a subring of T containing R and if Q is an ideal of B such that $Q \cap R = P$, then $B = R$.
(2) If $x \in T \setminus R$, then there exists $x' \in P$ with $xx' \in R \setminus P$.
(3) There is a valuation on T such that $R = \{x \in T : v(x) \geq 0\}$ and $P = \{x \in T : v(x) > 0\}$.

PROOF: (1) $\Rightarrow$ (2): By the Lying Over theorem [K, p. 29], R is integrally closed in T. There is nothing to prove if $R = T$. Assume that $R \subset T$ and choose $x \in T \setminus R$. Let $B = R[x]$ and $Q = PB$. Then Q is an ideal of B such that $Q \cap R \supset P$. Let $a \in (Q \cap R) \setminus P$ and write $a = p_o + p_1 x + \cdots + p_n x^n$ where $p_i \in P$ and n is minimal. Multiplying by p_n^{n-1} shows that xp_n is integral over R; and hence in R. If $xp_n \in P$, then $a = p_o + p_1 x + \cdots + (p_{n-1} + xp_n)x^{n-1}$, contradicting the minimality of n. Therefore p_n is the required element.

(2) $\Rightarrow$ (3): For $x \in T$, define $(P : x) = \{z \in T : zx \in P\}$. Write $x \,\theta\, y$ if and only if $(P : x) = (P : y)$. This is an equivalence relation on T. Let $v(x)$ denote the equivalence class of x, and let $G = \{v(x) : x \in T\} \setminus \{v(0)\}$. For $v(x), v(y) \in G$ define $v(x) + v(y) = v(xy)$, and define $v(x) > 0$ if $x \in P$. Finally, set $v(0) = \infty$, where $v(x) \leq \infty$ for all $v(x) \in G$. Under these definitions $G \cup \{\infty\}$ is an extended totally ordered abelian group,

$v : T \to G \cup \{\infty\}$ is a valuation, $R = \{x \in T : v(x) \geq 0\}$ and $P = \{x \in T : v(x) > 0\}$.

(3) $\Rightarrow$ (2) and (2) $\Rightarrow$ (1) are straightforward.

If the equivalent conditions of Theorem 5.1 hold, then (R, P) is called a *valuation pair* of T and R is called a *valuation ring* of T, or just a *valuation ring*. Valuation pairs will usually be denoted by (V, M).

Theorem 5.2 If $R \subseteq T$ are rings and P is a prime ideal of R, then there exists a valuation pair (V, M) of T such that $V \supseteq R$ and $M \cap R = P$.

PROOF: Use Zorn's lemma to find a pair (V, M) satisfying condition (1) of Theorem 5.1.

If the function v in the above definition of a valuation is *into*, but not necessarily *onto*, then v is a *paravaluation*. These maps play an important role in Section 9. If v is a paravaluation on T and if $R = \{x \in T : v(x) \geq 0\}$, then R is a ring. Such rings are called *paravaluation rings*.

Let R be a subring of T. A *dominated polynomial* over R is a polynomial $P = X_1^{n(1)} \cdots X_r^{n(r)} + \sum_\delta a_\delta X_1^{\delta(1)} \cdots X_r^{\delta(r)}$, where the X_i's are indeterminates, $a_\delta \in R$, and $(n(1), \ldots, (n(r)) > (\delta(1), \ldots, \delta(r))$ with the order given by the ordered product of r copies of the natural numbers. The ring R has the *dominated polynomial property*, if for each dominated polynomial $P = P(X_1, \ldots, X_r)$ over R and for each set $s_1, \ldots, s_r \in T \setminus R$, $P(s_1, \ldots, s_r) \neq 0$.

Lemma 5.3 Let $R \subseteq T$ be rings. If R is a valuation or a paravaluation subring of T, or if R has the dominated polynomial property with respect to T; then $T \setminus R$ is a multiplicatively closed set.

PROOF: If R is a paravaluation ring, use part (a) of the definition. Assume that R satisfies the dominated polynomial property. Let $s_1, s_2 \in T \setminus R$, and assume that $s_1 s_2 = a \in R$. Then $P = X_1 X_2 - a$ is a dominated polynomial over R, but $P(s_1, s_2) = 0$.

Lemma 5.4 Let R be a subring of T such that $T \setminus R$ is multiplicatively closed. Assume that $A = \{x \in T : xy = 0 \text{ for some } y \in T \setminus R\}$. Then:

(1) A is a common ideal of R and T.
(2) $R \subseteq T$ satisfies the dominated polynomial property if and only if $R/A \subseteq T/A$ does.
(3) R is a paravaluation ring of T if and only if R/A is a paravaluation ring of T/A.

PROOF: The proof of (1) is clear. If P is a polynomial over R, let $\bar{P}$ be the polynomial over R/A obtained by reducing the coefficients of P modulo A.

(2): Assume that $R \subseteq T$ has the dominated polynomial property, and let $\bar{P}$ be a dominated polynomial over R/A. Then P (some polynomial over R from which $\bar{P}$ is derived) is dominated over R. Choose $\bar{s}_1, \ldots, \bar{s}_r \in T/A \setminus R/A$, where $\bar{s}_i = s_i + A$. Then $s_1, \ldots, s_r \in T \setminus R$; so $P(s_1, \ldots, s_r) \neq 0$. If $\bar{P}(\bar{s}_1, \ldots, \bar{s}_r) = \bar{0}$ then $P(s_1, \ldots, s_r) = a \in A \subseteq R$. The new polynomial $P_1 = P - a$ is also dominated over R, but $P_1(s_1, \ldots, s_r) = 0$. Therefore $\bar{P}(\bar{s}_1, \ldots, \bar{s}_r) \neq \bar{0}$. The converse is easy.

(3): Assume that R is a paravaluation subring of T. Let v be the associated paravaluation. If $a \in A$, then $v(a) = \infty$; for there exists $y \in T \setminus R$ such that $ay = 0$, which implies that $v(a) + v(y) = v(ay) = \infty$.

If $v : T \to G \cup \{\infty\}$, define $\bar{v} : T/A \to G \cup \{\infty\}$ by $\bar{v}(t + A) = v(t)$. Then $\bar{v}$ is the required paravaluation map. (Use the fact that $a \in A$ implies $v(a) = \infty$ to get well-definedness.) The converse is straightforward.

Theorem 5.5 A ring R is a paravaluation subring of T if and only if R satisfies the dominated polynomial property with respect to T.

PROOF: ($\Rightarrow$): Let $v : T \to G \cup \{\infty\}$ be a paravaluation, and let R be its associated ring. Let $P = X_1^{n(1)} \cdots X_r^{n(r)} + \sum_\delta a_\delta X_1^{\delta(1)} \cdots X_r^{\delta(r)}$ be a dominated polynomial over R. If $s_1, \ldots, s_r \in T \setminus R$, then $P(s_1, \ldots, s_r) = s_1^{n(1)} \cdots s_r^{n(r)} + \sum_\delta a_\delta s_1^{\delta(1)} \cdots s_r^{\delta(r)}$. Consequently $v(P(s_1, \ldots, s_r)) = v(s_1^{n(1)} \cdots s_r^{n(r)}) < 0$, and therefore $P(s_1, \ldots, s_r) \neq 0$.

($\Leftarrow$): Let $\psi : T \to T/A$ be the canonical homomorphism. Then $\psi(R) = R'$ has the dominated polynomial property with respect to $\psi(T) = T'$, Lemma 5.4. We need to prove that R' is a paravaluation ring. Note that the multiplicatively closed set $S = T' \setminus R'$ consists entirely of regular elements of T'. Define $R'(S^{-1})$ to be the subring of T'_S generated by R' and $\{a/s : a \in R', s \in S\}$. Then $R'(S^{-1}) \cap T' = R'$; for if $a_o + \sum_{i=1}^n a_i/s_i = s_o \in S = T' \setminus R'$, then multiplying by $\prod_{i=1}^n s_i$ leads to a contradiction of the dominated polynomial property.

The set $P = \{x \in R : xy \in R \text{ for some } y \in T \setminus R\}$ is a prime ideal of R. Let Q be the ideal of $R'(S^{-1})$ generated by $\psi(P) = P'$ and $\{a/s : a \in R', s \in S\}$. If $Q = R'(S^{-1})$, then $1 = p' + \sum_{i=1}^n a_i/s_i$ with $p' \in P'$, $a_i \in R'$, and $s_i \in S$. There exists $s_o \in S$ such that $p's_o \in R'$; so $s_o = p's_o + \sum_{i=1}^n s_o a_i/s_i$, and again multiplying by $\prod_{i=1}^n s_i$ yields a contradiction.

Let M' be a prime ideal of $R'(S^{-1})$ such that $Q \subseteq M'$. Let (V, M) be a valuation pair of T'_S such that $V \supseteq R'(S^{-1})$ and $M \cap R'(S^{-1}) = M'$. Let v be the valuation associated with V. If v_o is the restriction of v to T', then it is certainly a paravaluation on T'. Suppose that $s \in T' \setminus R'$; then $1/s \in Q$, so $v(1/s) > 0$ which implies that $s \notin V$. Thus $\{x \in T' : v(x) \geq 0\} \subseteq R'$. On the other hand $R' \subseteq R'(S^{-1}) \subseteq V$, so $\{x \in T' : v_o(x) \geq 0\} = R'$.

Most of the time we are interested in valuations and paravaluations on total quotient rings. However there are times when it is more convenient to discuss these functions with respect to arbitrary rings. The reason for this can be seen from Lemma 5.4. If v is a paravaluation on a total quotient ring T, then $\bar{v}(\bar{x}) = v(x)$, where $\bar{x} = x+A$, is a paravaluation on T/A; but there is no guarantee that T/A is a total quotient ring. The times when valuations or paravaluations are defined on rings other than total quotient rings will be explicitly noted.

For an integral domain R, the five conditions of Theorem 5.1 and 5.5 are equivalent. They are also equivalent to the condition that the ideals of R are linearly ordered by inclusion—the usual definition of a valuation domain.

6. Prüfer Rings

Let R be a ring and N be a multiplicatively closed subset of R. In addition to the usual quotient ring R_N, there are other "quotient" rings of R that can be derived from N.

(a) $R_{(N)} = \{a/b \in T(R) : a, b \in R, b \in N,$ and b is regular$\}$. This is called the *regular quotient ring* of R with respect to N, and it is contained in $T(R)$.

(b) $R_{[N]} = \{z \in T(R) : zs \in R,$ for some $s \in N\}$. The ring $R_{[N]}$ is called the *large quotient ring* of R with respect to N. It contains $R_{(N)}$ and is contained in $T(R)$.

If S is an R-submodule of $T(R)$, then $[S]R_{[N]} = \{z \in T(R) : zs \in S,$ for some $s \in N\}$ is an $R_{[N]}$-submodule of $T(R)$. In particular, if I is an ideal of R, then $[I]R_{[N]}$ is an ideal of $R_{[N]}$. For a prime ideal P of R, write $R_{(P)}$ (resp., $R_{[P]}$) in place of $R_{(R\setminus P)}$ (resp., $R_{[R\setminus P]}$). Clearly $[P]R_{[P]}$ is a prime ideal of $R_{[P]}$. In the zero divisor case the rings $R_{(N)}$ and $R_{[N]}$, being overrings of R, possess some important properties that R_N does not have. The next result exhibits two such properties.

Theorem 6.1 Let R be a ring and S an R-submodule of $T(R)$. If $\{M_\lambda\}$ is the set of all maximal ideals of R and if $\{M_\sigma\}$ is the set of regular maximal ideals of R, then:

(1) $S = \cap SR_{(M_\lambda)} = \cap SR_{[M_\lambda]} = \cap[S]R_{[M_\lambda]}$.

(2) If S contains a regular element of R, then $S = \cap SR_{(M_\sigma)} = \cap SR_{[M_\sigma]} = \cap[S]R_{[M_\sigma]}$.

PROOF: (1): Clearly $S \subseteq \cap SR_{(M_\lambda)} \subseteq \cap SR_{[M_\lambda]} \subseteq \cap[S]R_{[M_\lambda]}$. If $x \in \cap[S]R_{[M_\lambda]}$, then $A = \{r \in R : rx \in S\}$ is an ideal of R. Since $x \in [S]R_{[M_\lambda]}$ there exists some $y \in R \setminus M_\lambda$ such that $xy \in S$. Thus $A \not\subseteq M_\lambda$. Since this is true for each λ, $A = R$ and $x = 1x \in S$.

(2): As above, it suffices to show that $\cap[S]R_{[M_\sigma]} \subseteq S$. If $x \in \cap[S]R_{[M_\sigma]}$, then $A = \{r \in R : rx \in S\}$ is an ideal of R. Also x is in $T(R)$, so $xy \in R \subseteq S$ for some regular element y in R. If b is a regular element of R contained in S, then $xyb \in S$; hence A is a regular ideal. Since $x \in [S]R_{[M_\sigma]}$, there exists some $z \in R \setminus M_\sigma$ such that $xz \in S$. Thus A is not contained in a regular maximal ideal of R. Therefore $A = R$ and $x \in S$.

If I is an ideal of R, let $I^{-1} = \{z \in T(R) : zI \subseteq R\}$. As in the integral domain case, define I to be *invertible* if $II^{-1} = R$. Invertible ideals are regular. (To see this suppose that I is invertible. Choose $a_i \in I$ and $b_i/d \in I^{-1}$ such that $1 = \sum a_i(b_i/d)$, where d is regular. Then $d = \sum a_i b_i \in I$.) The preliminaries are now finished. Define a *Prüfer ring* to be a ring for which each finitely generated regular ideal is invertible. A Prüfer ring where each regular finitely generated ideal is principal is called a *Bezout ring*. Let P be a prime ideal of R. If the image in R_P of every pair of ideals of R, one at least of which is regular, are totally ordered by inclusion, then (R, P) is said to have the *regular total order property*.

Theorem 6.2 The following conditions are equivalent for a ring R:

(1) R is a Prüfer ring.
(2) $(R_{[M]}, [M]R_{[M]})$ is a valuation pair for each maximal ideal M of R.
(3) Each overring of R is integrally closed in $T(R)$.
(4) If $IJ = IK$ where I, J, K are ideals of R and I is finitely generated and regular, then $J = K$.
(5) (R, M) has the regular total order property for every maximal ideal M.

The proof of this result is in Larsen and McCarthy's book [LM].

The integral domain case and the ring theory case diverge unexpectedly at this point. There exist valuation rings that are not Prüfer. If (V, P) is a valuation pair, P may not be the unique regular maximal ideal of V, and it may happen that P is not even a maximal ideal of V. See [LM] and our last chapter for examples. However, the Prüfer valuation rings have reasonable properties. Before giving their structure some preliminaries are needed.

For an ideal I of the ring R, define the *ideal transform* of I to be $S = \{x \in T(R) : xI^n \subseteq R, \text{ for some positive integer } n\}$. Clearly S is an overring of R, and if I is an invertible ideal of R, then $IS = S$.

Lemma 6.3 Let I be a finitely generated regular ideal in a Prüfer ring R. If $P \in \operatorname{Spec} R$, then $S \subseteq R_{[P]}$ if and only if $I \not\subseteq P$.

PROOF: ($\Leftarrow$): Let S be the ideal transform of I. If $x \in I \setminus P$, then $z \in S$ implies that $zx^n \in R$ for some positive integer n. Thus $z \in R_{[P]}$.

($\Rightarrow$): Assume that $I \subseteq P$. Since I is invertible, $S = IS = PS$. It is easy to see that $[P]R_{[P]} \cap R = P$, so $PR_{[P]} \subset R_{[P]}$. But $S = PS$ implies that $PR_{[P]} = R_{[P]}$, a contradiction.

Lemma 6.4 Let R be a Prüfer ring, and let M and N be regular prime ideals of R. Then $R_{[M]} \subseteq R_{[N]}$ if and only if $N \subseteq M$.

PROOF: ($\Leftarrow$): Clear.

($\Rightarrow$): Assume that $a \in N \setminus M$ and let b be a regular element of N. If S is the ideal transform of (a, b), then $S \subseteq R_{[M]}$ and $S \not\subseteq R_{[N]}$, a contradiction.

Theorem 6.5 For a ring R and a prime ideal P of R, the following conditions are equivalent:

(1) (R, P) is a Prüfer valuation pair.
(2) R is a Prüfer ring and P is the unique regular maximal ideal of R.
(3) R is a valuation ring and P is the unique regular maximal ideal of R.

PROOF: (1) $\Rightarrow$ (2): Use Lemma 6.4.

(2) $\Rightarrow$ (3): Since R is Prüfer and $R_{[P]} = R$ (Theorem 6.1), R is a valuation ring.

(3) $\Rightarrow$ (1): This follows from the facts that $R = R_{[P]}$ and $P = [P]R_{[P]}$.

7. Marot Rings

Define a ring R to be a *Marot ring* if each regular ideal of R is generated by its set of regular elements.

Theorem 7.1 The following conditions on a ring R are equivalent:

(1) R is a Marot ring.
(2) Every pair of elements $\{a, b\}$ in R with b regular has the property that the ideal (a, b) admits a finite system of regular elements as generators.

(3) Every regular R-module contained in $T(R)$ admits a system of regular elements as generators.

PROOF: (1) $\Rightarrow$ (2): Let $\{g_\alpha\}$ be the regular elements in $I = (a,b)$. Let $a = r_1 g_{\alpha_1} + \cdots + r_n g_{\alpha_n}$ and $b = s_1 g_{\beta_1} + \cdots + s_m g_{\beta_m}$ for appropriate g's, where $r_i, s_j \in R$. Then I is generated by the finite set $\{g_{\alpha_i}\} \cup \left\{g_{\beta_j}\right\}$.

(2) $\Rightarrow$ (3): Suppose g is a regular element in A, an R-submodule of $T(R)$. The union of a set of generators of the R-module $aR + gR$, as a varies over A, is a set of generators for A. Hence it suffices to find a set of regular generators for $aR + gR$. Let s be the common denominator of a and g. Then (as, gs) is a regular ideal of R, so it has a finite set of regular generators; say, $p_1, \ldots, p_t$. Then $\{p_1/s, \ldots, p_t/s\}$ is a regular generating set for $aR + gR$.

(3) $\Rightarrow$ (1): Clear.

A ring R has *few zero divisors* if $Z(R)$ is a finite union of prime ideals; and R is *additively regular* if for each $z \in T(R)$, there exists $u \in R$ such that $z + u$ is a regular element of $T(R)$. The extensiveness of Marot rings can be seen by the next four results.

Theorem 7.2 Consider the following four conditions on a ring R.

(1) R is a Noetherian ring.
(2) R has few zero divisors.
(3) R is an additively regular ring.
(4) R is a Marot ring.

Then, (1) $\Rightarrow$ (2) $\Rightarrow$ (3) $\Rightarrow$ (4).

PROOF: (1) $\Rightarrow$ (2): Use [K, p. 55].

(2) $\Rightarrow$ (3): Let $Z(R) = \cup_{i=1}^{n} P_i$, where the P_i are prime ideals. We may assume that there are no containment relations among the P_i. If $a/b \in T(R)$ with b regular, assume that $a \in \cap_{i=1}^{s} P_i \setminus$

$\cup_{i=s+1}^{n} P_i$. Choose $u \in \cap_{i=s+1}^{n} P_i \setminus \cup_{i=1}^{s} P_i$. Then $a/b + u$ is a regular element.

(3) $\Rightarrow$ (4): Let I be an ideal containing a regular element b. For each $a \in I$, there exists $u \in R$ such that $a + ub = r_a$ is regular. Then the ideal generated by $\{r_a\} \cup \{b\}$ is I.

The proof of the last theorem shows that if I is a regular ideal generated by n elements in an additively regular ring, then I requires at most $n+1$ regular generators. None of the implications of Theorem 7.2 are reversible. (2) $\not\Rightarrow$ (1) is easy. (3) $\not\Rightarrow$ (2) follows because a direct product of a family of rings $\{R_\alpha\}$ is additively regular if and only if each R_α is additively regular. That (4) $\not\Rightarrow$ (3) is more difficult and will be given in the section entitled Examples.

Corollary 7.3 Each overring of a Marot ring is a Marot ring.

PROOF: Use Theorem 7.1.

Theorem 7.4 Let R be a ring with 0-dimensional total quotient ring, then R is additively regular (and hence Marot).

PROOF: If $T(R)$ is reduced, then it is a von Neumann regular ring. Let $a/b \in T(R)$. The object is to find some $u \in R$ such that $a/b + u$ is regular; that is, such that $a + bu$ is regular. Let $\operatorname{Min} R = \{P_i\} \cup \{Q_j\}$ where $a \in \cap P_i \setminus \cup Q_j$. Consider a as an element in $T(R)$ and choose $c \in \cap Q_j T(R) \setminus \cup P_i T(R)$. Write $c = u/c_1$ where $u, c_1 \in R$ and c_1 is regular. Then $u \in \cap Q_j \setminus \cup P_i$. Since $a + bu$ is not contained in any minimal prime ideal of R, it is regular.

Assume that $T(R)$ has a nonzero nilradical N. Then

$$R/(R \cap N) \subseteq T(R)/N \subseteq T(R/(R \cap N))$$

Since $T(R)/N$ is a von Neumann regular ring, $T(R)/N = T(R/(R \cap N))$. By the first part of the proof, $R/(N \cap R)$ is additively regular. Thus if the coset $z + N$ is in $T(R)/N$, there exists a $u \in R$ such that $(z+N)+(u+N) = b+N$ is in $T(R)/N$. Then $z+u = b+n$ where $n \in N$. If $P \in \operatorname{Spec} T(R)$, then $b+N \notin P/N$ which implies that $b \notin P$. It follows that $b+n$ is a regular element of $T(R)$. This completes the proof.

Theorem 7.5 Let R be a graded ring containing a regular homogeneous element of positive degree. Then R is an additively regular ring.

PROOF: Let a and b be in R with b regular. Let $a = a_m + \cdots + a_n$ and $b = b_h + \cdots + b_k$ be the homogeneous representations of a and b. Say that x is a homogeneous regular element of R of degree $t > 0$. Choose a positive integer s such that $st + h > n$. If $a + x^s b$ is a zero divisor, then there is a nonzero homogeneous element c such that $c(a + x^s b) = 0$. Thus $cx^s b = 0$, which cannot happen since $x^s b$ is regular.

An important property of Marot rings is that they localize nicely. This is the content of the next theorem.

Theorem 7.6 Let R be a Marot ring. If $P \in \operatorname{Spec} R$, then $R_{[P]} = R_{(P)}$ and $PR_{(P)} = [P]R_{[P]}$.

PROOF: If P consists entirely of zero divisors, then $R_{(P)} = T(R) = R_{[P]}$. Consider the case when P is a regular prime ideal of R. Certainly $R_{(P)} \subseteq R_{[P]}$. If $x \in R_{[P]}$, there exists $z \in R \setminus P$ such that $zx \in R$. But x is in $T(R)$, so there exists a regular element b in R such that $bx \in R$. The ideal $I = (b, z)$ is regular and $Ix \subseteq R$. Since $I \nsubseteq P$, the Marot property implies the existence of a regular element $u \in I \setminus P$ such that $ux \in R$. Therefore $R_{(P)} = R_{[P]}$.

For the second part notice that $PR_{(P)} \subseteq [P]R_{[P]} \subset R_{(P)}$ and $PR_{(P)}$ is a maximal ideal; hence $PR_{(P)} = [P]R_{[P]}$.

The Marot hypothesis greatly simplifies the structure of valuation rings.

Theorem 7.7 Let V be a Marot ring. Then the following conditions are equivalent:

(1) V is a Prüfer valuation ring.
(2) V is a valuation ring.
(3) V is a paravaluation ring.
(4) For each regular $x \in T(V)$, x or x^{-1} is in V.

PROOF: The only thing that needs to be proved is that (4) $\Rightarrow$ (1). Let M and N be regular maximal ideals of V and let $\{r_\alpha\}$ be the set of regular nonunits of V. Then $\{r_\alpha\}$ is contained in both M and N. Thus $M = N$. In view of Theorem 6.5, it is enough to prove that (V, M) is a valuation ring of $T(V)$. Choose $x = t/s \in T(V) \setminus V$, where $t, s \in V$ and s is regular. If t is regular, then $s/t \in M$ and hence $xs/t = 1 \in V \setminus M$. On the other hand, if t is a zero divisor, then the regular ideal (t, s) of V can be written as $(t, s) = (a)$ where a is regular. Writing $a = \alpha t + \beta s$, $t = t'a$, and $s = s'a$ where all elements are in V, we have $1 = \alpha t' + \beta s'$; hence $(s', t') = V$. If $s' \in V \setminus M$, then $x \in V$, contrary to the assumption. Therefore $t' \in V \setminus M$ and $s' \in M$. Observe that $ts'/s = t' \in V \setminus M$.

Corollary 7.8 Overrings of Marot valuation rings are valuation rings.

All Prüfer valuation rings are not Marot rings; see the Example section. However the Marot rings in this set can be identified.

Theorem 7.9 Let V be a Prüfer valuation ring with corresponding valuation v and value group G. The regular elements of V map onto the positive elements of G if and only if V is a Marot ring.

PROOF: ($\Rightarrow$): It is sufficient to consider an ideal (a, b) of V, with b regular, and show that it is generated by regular elements. If a is regular or if $v(a) \geq v(b)$, then the result follows easily. Assume that a is a zero divisor such that $v(a) < v(b)$. Let $v(r) =$

$v(a)$ where r is a regular element. Then $a = rc$, $c \in Z(V)$ and $v(c) = 0$. Also, $b = rd$ where d is regular and $v(d) > 0$. The ideal (c, d) is regular and c is not contained in M, the unique regular maximal ideal of V, Theorem 6.5. Since d does not belong to any maximal ideal contained in $Z(V)$, $(c, d) = V$. If $x, y \in V$ such that $xc + yd = 1$, then $xa + yb = r$. Therefore $(a, b) = (r)$.

($\Leftarrow$): Let α be a positive (finite) element in G and let $x \in V$ such that $v(x) = \alpha$. Choose $a/b \in T(V)$ with a and b in V and b regular such that $-v(x) = v(a/b)$. Then $v(x) = v(b) - v(a) \leq v(b)$. The ideal (x, b) of V is principal and is generated by a regular element r. Clearly $v(x) \geq v(r)$. On the other hand, $r = sx + tb$ for appropriate $s, t \in V$. Thus $v(r) \geq \min\{v(sx), v(tb)\} \geq v(x)$. Therefore $v(r) = v(x)$.

For our next result we need two definitions. An ideal Q of R is *prime* (resp., *primary*) *for its regular elements* if whenever x and y are regular elements of R such that $xy \in Q$, then $x \in Q$ or $y \in Q$ (resp., then $x \in Q$ or $y \in \operatorname{Rad} Q$).

Theorem 7.10 Let R be a Marot ring. Then a regular ideal Q of R is prime (resp., primary) if and only if Q is prime (resp., primary) for its regular elements.

PROOF: Only the proof for primary ideals is given. If a regular ideal Q is primary, then it is certainly primary for its regular elements.

For the converse assume that Q is a regular ideal which is not primary. There are elements $x, y \in R$ such that $xy \in Q$, $x \notin Q$, and $y \notin \operatorname{Rad} Q$. Choose a regular element $z \in (Q, x) \setminus Q$. Since (Q, y) is generated by its regular elements, $y = a_1t_1 + \cdots + a_nt_n$ where each t_i is a regular element in (Q, y). Since $y \notin \operatorname{Rad} Q$, some $t_i \notin \operatorname{Rad} Q$. Thus $zt_i \in Q$, showing that Q is not primary for its regular elements.

8. Krull Rings I

In this section all rings are assumed to have the Marot property. Hence each valuation ring will have a unique regular maximal ideal.

Let V be a valuation ring with corresponding valuation v and value group G. Then V (resp., v) is called a *discrete rank one valuation ring* (resp., *discrete rank one valuation*) if G is isomorphic to the group of integers.

Lemma 8.1 Let (V, M) be a discrete rank one valuation ring. Then:

(1) M is the unique regular prime ideal of V and there exists a regular $x \in M$ such that $M = (x)$.

(2) V is a maximal subring of $T(V)$.

PROOF: (1): Choose x regular in M such that $v(x) = 1$ (Theorem 7.9). If r is another regular nonunit in V, then $r \in M$ (Theorem 6.5) and $v(r) = n > 0$. Thus x^n and r are associates in V. This is enough to show the uniqueness of M. It is easy to see that x generates M.

(2): Let $V \subseteq W \subset T(V)$. The Marot property implies that W is a Prüfer valuation ring with unique maximal ideal P. Thus $P \subseteq M$, so $P = M$. This shows that $V = W$.

A ring R is a *Krull ring* if either $R = T(R)$ or if there exists a family $\{v_i\}$ of discrete rank one valuations such that:

(a) R is the intersection of the corresponding valuation rings $\{V_i\}$.

(b) For each regular $x \in T(R)$, $v_i(x) = 0$ for all but a finite number of i.

Part (b) merely says that each regular element of $T(R)$ is a unit in all but finitely many V_i. We have included $R = T(R)$ as a Krull ring for technical reasons. (In this case the defining family of valuation overrings is empty.) In the proofs that follow we rarely need to refer to this case, since the results will be clear for total quotient rings.

Theorem 8.2 If S is a multiplicatively closed subset of a Krull ring R, then $R_{(S)}$ is a Krull ring.

PROOF: Write $R = \cap V_i$ where $\{V_i\}$ are the defining valuation rings for R. It is clear that $R_{(S)} \subseteq \cap(V_{i(S)})$. If z is a regular element of $\cap(V_{i(S)})$ and if $V_{i_1}, \ldots, V_{i_m}$ are the valuation rings for which z is not a unit, then there is some $s \in S$ such that $zs \in V_{i_j}, j = 1, \ldots, m$. Thus $zs \in \cap V_i = R$, and hence $z \in R_{(S)}$. By Theorem 7.1, $R_{(S)} = \cap(V_{i(S)})$. Lemma 8.1 implies that for each i, $V_{i(S)} = V_i$ or $V_{i(S)} = T(R)$. Deleting the $V_{i(S)}$ that equal $T(R)$, we have $R_{(S)} = \cap V_j$ where $V_{j(S)} = V_j$.

We indicate how to extend the theory of divisor class groups from integral domains to rings with zero divisors. The proofs for integral domains appear in Section 34 of [G]. These proofs may be generalized in a straightforward, but tedious manner. We will not present the proofs of these generalizations.

For a ring R, let $\mathcal{F}(R)$ be the set of *fractional ideals* of R in the sense of Kaplansky [K]—that is, $A \in \mathcal{F}(R)$ in case A is an R-submodule of $T(R)$. Define $\mathcal{F}^*(R)$ to be the subset of $\mathcal{F}(R)$ consisting of all regular R-modules A of $T(R)$ such that there exists some regular $d \in R$ so that $dA \subseteq R$. For $A \in \mathcal{F}(R)$ let $A^{-1} = (R : A)$, where $(R : A) = \{x \in T(R): xA \subseteq R\}$. Denote $(A^{-1})^{-1}$ by A_v. A fractional ideal A is *divisorial* if $A = A_v$. By [G, p. 26], $A \in \mathcal{F}^*(R)$ implies that $A^{-1} \in \mathcal{F}^*(R)$. (Note: Our $\mathcal{F}^*(R)$ is not the same as the $\mathcal{F}^*(R)$ defined in [G].) Let $A, B \in \mathcal{F}^*(R)$. Define $A \sim B$ in case $A_v = B_v$. This is an equivalence relation on $\mathcal{F}^*(R)$ and we denote the equivalence class of A by div A. Let $\mathcal{D}(R)$ be the set of all div A and define a binary operation on $\mathcal{D}(R)$ by div $A \oplus$ div B = div AB. Then $(\mathcal{D}(R), \oplus)$ becomes an abelian semigroup with zero element div R. There is a natural ordering on $\mathcal{D}(R)$ compatible with $\oplus$ given by div $A \geq$ div B if and only if $A_v \subseteq B_v$. We call div A the *divisor class* of A. If $\mathcal{D}(R)$ is a group, it is called the *divisor class group* or R. If $A \in \mathcal{F}^*(R)$ write $(A : A)$ for $(A :_{T(R)} A)$.

Theorem 8.3 If $A \in \mathcal{F}^*(R)$, then A_v is the intersection of the set of principal fractional ideals of R containing A.

When R is a Marot ring (as it is in this section), the proof of Theorem 8.3 is identical to the proof for integral domains [G, p. 417]. However, as we shall see in the last chapter, when the Marot property is deleted, the theorem is no longer true.

Theorem 8.4 If the preceding notation is assumed, then the following conditions are equivalent:

(1) $\mathcal{D}(R)$ is a group.
(2) $(A : A) = R$ for each divisorial ideal $A \in \mathcal{F}^*(R)$.
(3) $(A : A) = R$ for each $A \in \mathcal{F}^*(R)$.
(4) R is completely integrally closed.

PROOF: Modify [G, p. 421].

For each $A \in \mathcal{F}^*(R)$ and for each i, $\inf\{v_i(a) : a \in A\}$ exists, since the value of a regular element is finite. For each i, denote this infinum by $v_i(A)$.

Lemma 8.5 Let R be a Krull ring with a defining family $\{v_i\}$ of discrete rank one valuations on $T(R)$.

(1) Let $A, B \in \mathcal{F}^*(R)$ where B is divisorial. Then $A \subseteq B$ if and only if $v_i(A) \geq v_i(B)$ for each i.
(2) For each $A \in \mathcal{F}^*(R), v_i(A) = 0$ for all but finitely many i.

PROOF: (1): Assume that $v_i(A) \geq v_i(B)$ for each i. Let $x \in A$ and $B = \cap y_\lambda R$ where $y_\lambda R$ are the principal fractional ideals of R containing B. If i and λ are fixed, then $v_i(x) \geq v_i(B) \geq v_i(y_\lambda)$ implies that $x \in y_\lambda V_i$. This holds for each i, hence $x \in \cap_i y_\lambda V_i = y_\lambda(\cap_i V_i) = y_\lambda R$. Therefore $x \in \cap y_\lambda R = B$. The converse is trivial.

(2): If $A \in \mathcal{F}^*(R)$, there exists a regular $d \in R$ such that $dA \subseteq R$. Choose a regular element $x \in dA$, then $(x) \subseteq dA \subseteq (x^{-1})R$. Therefore $v_i(x) \geq v_i(dA) \geq v_i(x^{-1})$ for each i. Apply part (b) of the definition of a Krull ring to complete the proof.

Theorem 8.6 Let R be a Krull ring. Then R is completely integrally closed and the ascending chain condition holds for divisorial ideals of R.

PROOF: We may assume that $R \neq T(R)$. Let $\{v_i\}$ be a defining family of discrete rank one valuations for R. Choose $x \in T(R)$ such that for some regular element d of R, $dx^n \in R$ for each positive integer n. If $x \notin R$, then there is some k such that $v_k(x) < 0$. Consequently $v_k(dx^n) = v_k(d) + nv_k(x) < 0$ for an appropriate n, a contradiction. Therefore R is completely integrally closed. The ascending chain condition for divisorial ideals is a consequence of Lemma 8.5.

A prime ideal P is called a *minimal regular prime* if $Q \subset P$ and Q is a prime ideal, imply that Q consists of zero divisors. The next result that we are aiming for is that if P is a minimal regular prime ideal of R, then $R_{(P)}$ is a discrete rank one valuation ring.

Lemma 8.7 Let P be a prime ideal of R. Then $R_{(P)}$ is equal to the R-module H generated by $\{x \in T(R) : x \text{ regular}, Rx^{-1} \cap R \not\subseteq P\}$.

PROOF: Suppose that $x = r/s$ is a regular element in $R_{(P)}$, $s \in R \setminus P$. Then $s \in Rx^{-1} \cap (R \setminus P)$. On the other hand let $x \in T(R)$ be regular such that $Rx^{-1} \cap R \not\subseteq P$. Choose $z = r/x \in$

$Rx^{-1} \cap (R \setminus P)$. Then $zx = r \in R$, so $x \in R_{[P]} = R_{(P)}$. By the Marot property $R_{(P)} = H$.

Lemma 8.8 If $R_{(P)}$, $P \in \operatorname{Spec} R$, is not a valuation ring, then there exists a regular $x \in T(R)$ such that $P^{-1} \subseteq (Rx \cap R : Rx \cap R)$.

PROOF: Choose a regular element $x \in T(R)$ such that $x, x^{-1} \notin R_{(P)}$. Then $(Rx \cap R) + (Rx^{-1} \cap R) \subseteq P$ (Lemma 8.7), and hence $P^{-1} \subseteq (R\colon Rx \cap R) \cap (R\colon Rx^{-1} \cap R)$. It remains to prove that $(R\colon Rx \cap R) \cap (R\colon Rx^{-1} \cap R) = (Rx \cap R : Rx \cap R)$. If $z(Rx \cap R) \subseteq Rx \cap R$, then $z \in (R\colon Rx \cap R)$. Furthermore $z(Rx \cap R)x^{-1} \subseteq (Rx \cap R)x^{-1}$, so $z(R \cap Rx^{-1}) \subseteq (R \cap Rx^{-1})$. Therefore $z \in (R\colon Rx \cap R) \cap (R : Rx^{-1} \cap R)$. For the opposite inclusion assume that $y(Rx \cap R) \subseteq R$ and $y(Rx^{-1} \cap R) \subseteq R$. Then $y(Rx^{-1} \cap R)x = y(Rx \cap R) \subseteq Rx \cap R$.

Lemma 8.9 If P is a regular prime ideal in a Krull ring R such that $P^{-1} \supset R$, then $R_{(P)}$ is a discrete rank one valuation ring.

PROOF: If $R_{(P)}$ is not a valuation ring, by Lemma 8.8 there exists a regular element x in $T(R)$ such that if $I = Rx \cap R$, then $P^{-1} \subseteq (I : I) = R$, a contradiction. If there exists a regular prime ideal Q of R such that $Q \subset P$, choose $y \in P \setminus Q$ with y regular. Then $yP^{-1} \subseteq R$ implies $yP^{-1}Q \subseteq Q$, and thus $P^{-1}Q \subseteq Q$. Therefore $P^{-1} \subseteq (Q : Q) = R$, again a contradiction. This proves that $R_{(P)}$ is a valuation ring and $PR_{(P)}$ is its unique regular prime ideal.

We claim that $PR_{(P)}$ is a principal ideal. Choose a regular element $x \in P^{-1} \setminus R$. Then $x \notin (P : P)$. Thus $xP \subseteq R$ and consequently $xPR_{(P)} \subseteq R_{(P)}$. Suppose that $xPR_{(P)} \subseteq PR_{(P)}$; then $xP \subseteq PR_{(P)} \cap R = P$, a contradiction. Thus $xPR_{(P)} = R_{(P)}$. This proves that $PR_{(P)}$ is an invertible ideal in the valuation ring $R_{(P)}$; hence $PR_{(P)} = aR_{(P)}$ is principal.

If b is a regular nonunit in $R_{(P)}$, then $\operatorname{Rad}(b) = P$. Thus (b) is $PR_{(P)}$-primary and therefore $(b) = P^n R_{(P)}$ for some positive integer n [G, p. 74]. For $x \in R_{(P)}$, define

$$v(x) = \begin{cases} 0 & \text{if } x \in R_{(P)} \setminus PR_{(P)} \\ n & \text{if } x \in P^n R_{(P)} \setminus P^{n+1} R_{(P)} \\ \infty & \text{if } x \in \cap P^n R_{(P)} \end{cases}$$

Then $v(xy) = v(x) + v(y)$ for all $x, y \in R_{(P)}$ [G, p. 74], and clearly $v(x+y) \geq \min\{v(x), v(y)\}$ for all $x, y \in R_{(P)}$. Since each regular element in $R_{(P)}$ has finite v-value, v may be extended to a valuation on $T(R_{(P)})$. Clearly $R_{(P)} = \{z \in T(R_{(P)}) : v(z) \geq 0\}$, and $PR_{(P)} = \{z \in T(R_{(P)}) : v(z) > 0\}$. We have proved that $(R_{(P)}, PR_{(P)})$ is a discrete rank one valuation ring.

Let $A, B \in \mathcal{F}^*(R)$ where A is a divisorial ideal. Since R is a Marot ring, $(A : B) = \cap A(1/b)$ where b varies over the regular elements of B. Since each $A(1/b)$ is a divisorial ideal, it follows that $(A : B)$ is also a divisorial ideal.

Theorem 8.10 Let R be a Krull ring and $\mathcal{H}$ the set of minimal regular prime ideals of R. Then:

(1) P is a maximal divisorial ideal of R if and only if $P \in \mathcal{H}$.
(2) $R_{(P)}$ is a rank one discrete valuation ring for each $P \in \mathcal{H}$.
(3) $R = \cap\{R_{(P)} : P \in \mathcal{H}\}$.

PROOF: (1) ($\Leftarrow$): Choose a regular element $z \in P$. Then $w = z^{-1} \notin R_{(P)}$; so $Rw^{-1} \cap R = Rz \cap R = (z)$ is a divisorial ideal in P of the form $Rw^{-1} \cap R$. By Theorem 8.6 choose a maximal divisorial ideal in P of the form $Rx^{-1} \cap R$. Call this ideal I. We prove that I is a prime ideal (and hence $I = P$). Assume that $a, b \in R$ such that $ab \in I$. In view of Theorem 7.10 we may assume that a and b are regular elements. It is clear that $R(xa)^{-1} \cap R \supseteq I$.

The first case to consider is when $R(xa)^{-1} \cap R \subseteq P$. Then $R(xa)^{-1} \cap R = I$. Hence $ba \in Rx^{-1} \cap R$, so $b = r/xa$ with $r \in R$, and therefore $b \in R(xa)^{-1} \cap R = I$.

The second case is when $R(xa)^{-1} \cap R \not\subseteq P$. Then $xa \in R_{(P)}$, Lemma 8.7. Thus $cxa \in R$ for some regular $c \in R \setminus P$; or equivalently, $ca \in Rx^{-1} \cap R = I$. An easy argument shows that $c(R(xc)^{-1} \cap R) \subseteq R(x^{-1}) \cap R \subseteq P$. Thus $R(xc)^{-1} \cap R \subseteq P$ and as in the first case $R(xc)^{-1} \cap R = I$. Therefore $a \in I$.

It remains to show that P is a maximal divisorial ideal of R. First assume that Q is a maximal divisorial integral ideal of R. We claim that Q is a prime ideal. Let I, J be ideals of R such that $IJ \subseteq Q$. By replacing I with $I + Q$ and J with $J + Q$ we may assume that $Q \subseteq I$ and $Q \subseteq J$. Suppose that $I \not\subseteq Q$, then $Q \subset I \subseteq (Q : J)$. But $(Q : J)$ is a divisorial ideal of R that properly contains Q. Thus $(Q : J) = R$ and so $J \subseteq Q$. This proves the claim.

Now suppose that Q is a maximal divisorial ideal of R that properly contains P. Then $T(R) \supset R_{(P)} \supset R_{(Q)}$ and $R_{(Q)}$ is a discrete rank one valuation ring, Lemma 8.9. We arrive at a contradiction by noting that $R_{(Q)}$ is not a maximal subring of $T(R)$, Lemma 8.1. Therefore P is a maximal divisorial ideal of R.

(1) ($\Rightarrow$): The argument given above shows that a maximal divisorial ideal is prime. Lemma 8.9 shows that it must be in $\mathcal{H}$.

(2): Lemma 8.9.

(3): Choose a regular element $x \in \cap R_{(P)}$. Then for each $P \in \mathcal{H}, Rx^{-1} \cap R \not\subseteq P$ (Lemma 8.7). Hence $Rx^{-1} \cap R$ is a divisorial ideal of R not contained in a maximal divisorial ideal; that is, $Rx^{-1} \cap R = R$. Therefore $x \in R$ and $R = \cap R_{(P)}$.

Valuation overrings of a Krull ring R of the form $R_{(Q)}, Q \in \operatorname{Spec} R$ are called *essential.*

Lemma 8.11 Let R be a Krull ring with a defining family of valuation rings $\{V_i\}$. If V is an essential valuation overring of R, then $V = V_i$ for some i. In particular V is a discrete rank one valuation ring.

PROOF: Let $V = R_{(Q)}$ for some regular prime ideal Q of R. Then $V = \cap V_j$ for some subset $\{V_j\}$ of $\{V_i\}$. Since V is a Marot valuation ring, its set of regular prime ideals are linearly ordered by inclusion. If V_{j_1} and V_{j_2} are two members of $\{V_j\}$, M_{j_1} and M_{j_2} are their respective unique regular maximal ideals, then $M_{j_1} \cap V = M_{j_1}$ and $M_{j_2} \cap V = M_{j_2}$. Suppose that $M_{j_1} \subseteq M_{j_2}$, then $V_{j_1} \supseteq V_{j_2}$. (For suppose that x is a regular element not in V_{j_1}; then $1/x \in M_{j_1} \subseteq M_{j_2}$, which implies that $x \notin V_{j_2}$.) This proves that $V_{j_1} = V_{j_2}$, and so $V = V_{j_1}$.

Corollary 8.12 If R is a Krull ring with a defining family of valuation rings $\{V_i\}$ and if P is a minimal regular prime ideal of R, then $R_{(P)}$ is one of the V_i.

Corollary 8.13 Let R be a Krull ring and Let $\{P_i\}$ be the minimal regular prime ideals of R. Then $\{R_{(P_i)}\}$ is the unique defining family of essential discrete rank one valuation rings for R.

Lemma 8.14 Let R be a reduced ring with a finite number of minimal prime ideals $\{Z_i\}_{i=1}^{g}$. If K_i is the quotient field of R/Z_i, then:

(1) $\{Z_i\}$ is precisely the set of prime ideals of R contained in $Z(R)$.
(2) $T(R) \cong \sum K_i$ (direct).
(3) $R' \cong \sum(R/Z_i)'$ (direct).

PROOF: (1): Let P be a nonminimal prime ideal of R. Choose $x \in P \setminus \cup Z_i$. Then $xy = 0$ for some $y \in R$ implies that $y \in \cap Z_i = (0)$. Hence P is a regular prime ideal.

(2): Let $Z_i^e = Z_iT(R)$, then $\{Z_i^e\} = \operatorname{Spec} T(R)$. It is easy to see that $K_i \cong T(R)/Z_i^e$, and it is well known that $T(R) \cong \sum T(R)/Z_i^e$ (direct).

(3): The mapping $\varphi : R \to \sum R/Z_i$ defined by $\varphi(r) = (r + Z_1, ..., r + Z_g)$ is a monomorphism. Hence $R \cong \varphi(R) \subseteq \sum R/Z_i \subseteq \sum(R/Z_i)'$. Suppose that $y = (0, \ldots, r + Z_i, \ldots, 0)$, $r + Z_i \in R/Z_i$. Then $y^2 - \varphi(r)y = 0$. It follows that $\sum R/Z_i \subseteq \varphi(R)'$. Therefore $R' \cong \varphi(R)' = \sum(R/Z_i)'$.

Corollary 8.15 Let R be a reduced ring with a finite number of minimal prime ideals $\{Z_i\}_{i=1}^g$. Let $\{Z_i'\}_{i=1}^g$ be the minimal prime ideals of R'. Then $(R/Z_i)' \cong R'/Z_i'$, for each i.

PROOF: Write $R' \cong (R/Z_1)' \oplus \cdots \oplus (R/Z_i)' \oplus \cdots \oplus (R/Z_g)'$. Then $Z_i' \cong (R/Z_1)' \oplus \cdots \oplus (0) \oplus \cdots \oplus (R/Z_g)'$, and therefore $R_i'/Z_i' \cong (R/Z_i)'$.

If D is a Krull domain, then the polynomial ring $D[X]$ is also Krull. This is not necessarily true for rings with zero divisors. For an example, suppose that R is a ring with a nontrivial nilpotent element c. By McCoy's theorem X is a regular element in $R[X]$. Hence c/X is integral over $R[X]$, but not in $R[X]$. Therefore, if R is a Krull ring with nonzero nilradical, then $R[X]$ is not a Krull ring.

We have just seen that X is always a regular element in $R[X]$. However, X need not be irreducible, as the next example shows. Let $R = Z \oplus Z$ where Z is the ring of integers. In $R[X]$, $X = ((1,0) + (0,1)X)((0,1) + (1,0)X)$.

Our next result characterizes those rings for which $R[X]$ is a Krull ring. In the proof of this result we write $R = D_1 \oplus \cdots \oplus D_n$ where each D_i is an integral domain. Some (or all) of the D_i may be fields. If so they are automatically Krull domains with empty families of defining valuation rings.

Theorem 8.16 If R is a ring, then $R[X]$ is a Krull ring if and only if R is a finite direct sum of Krull domains.

PROOF: Assume that $R[X]$ is a Krull ring. Since $R[X]$ is integrally closed, R is integrally closed and reduced. Since $R \cong R[X]/(X)$, (X) is a radical ideal. (If $(a_0+a_1X+\cdots+a_tX^t)^n \in (X)$ then $a_0^n = 0$, so $a_0 = 0$.) By Corollary 8.13 the unique defining family of essential valuation rings for $R[X]$ is $\{R[X]_{(\mathcal{P}_i)}\}$ where $\{\mathcal{P}_i\}$ is the set of minimal regular prime ideals of $R[X]$. Thus $(X) = X(\cap R[X]_{(\mathcal{P}_i)}) = \cap XR[X]_{(\mathcal{P}_i)} = \cap(XR[X]_{(\mathcal{P}_i)} \cap R[X])$. By the definition of Krull ring there are only finitely many i such that $X(R[X]_{(\mathcal{P}_i)}) \neq R[X]_{(\mathcal{P}_i)}$. Suppose that $X \in \mathcal{P}_1, \ldots, \mathcal{P}_n$ and no other $\mathcal{P}_i$. Then $(X) = \cap_{i=1}^n(XR[X]_{(\mathcal{P}_i)} \cap R[X])$. Since $(X) = \text{Rad}(X)$, $(X) = \cap_{i=1}^n \mathcal{P}_i$.

Since $R \cong R[X]/(X)$, the zero element in R has a primary decomposition in R; say, $(0) = Z_1 \cap \cdots \cap Z_n$. Thus $\{Z_1, \ldots, Z_n\}$ is the complete set of minimal prime ideals of R. By Lemma 8.14 $R = \sum(R/Z_i)'$ (direct), and by Corollary 8.15 $R = \sum R/Z_i$ (direct). Letting $D_i = R/Z_i$ and K_i = quotient field of D_i, we have $R = D_1 \oplus \cdots \oplus D_n$ where each D_i is an integrally closed domain, some D_i may be fields, and $T(R) = K_1 \oplus \cdots \oplus K_n$. It remains to show that each D_i is a Krull domain. We concentrate on D_1. Let $\{V_\beta\}$ be the defining family of discrete rank one valuation rings for R. By parts (1) and (2) of Lemma 8.1, each V_β has the form $V_\beta = K_1 \oplus \cdots \oplus K_{i-1} \oplus V_{\beta_i} \oplus K_{i+1} \oplus \cdots \oplus K_n$, where V_{β_i} is a discrete rank one valuation domain between D_i and K_i. Let $\{V_\lambda\}$ be the subset of $\{V_\beta\}$ where each V_λ has the form $V_\lambda = V_{\lambda_1} \oplus K_2 \oplus \cdots \oplus K_n$. It is an easy matter to show that D_1 is a Krull domain with defining family $\{V_{\lambda_1}\}$.

Conversely assume that $R = D_1 \oplus \cdots \oplus D_n$ is a direct sum of Krull domains, some of which may be fields. Then $R[X] = D_1[X] \oplus \cdots \oplus D_n[X]$ where each $D_i[X]$ is a Krull domain. Write $D_i[X] = \cap_j V_{i_j}$, where $\{V_{i_j}\}$ is the defining family of discrete rank one valuation domains for $D_i[X]$. If $K_i(X)$ is the quotient field of $D_i[X]$, define

$$W_{i_j} = K_1(X) \oplus \cdots \oplus K_{i-1}(X) \oplus V_{i_j} \oplus K_{i+1}(X) \oplus \cdots \oplus K_n(X)$$

The value group of v_{i_j} is the additive group of integers Z. The mapping

$$w_{i_j} : K_1(X) \oplus \cdots \oplus K_i(X) \oplus \cdots \oplus K_n(X) \longrightarrow Z \cup \{\infty\}$$

given by $w_{i_j}((a_1, \ldots, a_i, \ldots, a_n)) = v_{i_j}(a_i)$ is a discrete rank one valuation on $T(R[X])$, and $W_{i_j} = \{x \in T(R[X]) : w_{i_j}(x) \geq 0\}$. It follows that $R[X] = \cap W_{i_j}$, and for each regular $x \in T(R[X])$, $w_{i_j}(x) = 0$ for all but a finite number of w_{i_j}.

If P is a regular prime ideal of a Marot ring R, then $PR_{(P)}$ is the unique regular maximal ideal of $R_{(P)}$. Let n be a positive integer. Then $P^n R_{(P)} \cap R$ is a P-primary ideal containing P^n. But the symbolic power of P^n, $P^{(n)} = P^n R_P \cap R$, is a minimal P-primary ideal containing P^n. On the other hand $P^n R_{(P)} \cap R \subseteq P^{(n)} R_{(P)} \cap R = P^{(n)}$. Therefore $P^{(n)} = P^n R_{(P)} \cap R$.

We sharpen the notion of primary decomposition of regular principal ideals that was used in the preceding proof of Theorem 8.16.

Theorem 8.17 If R is a Krull ring and d is a regular nonunit in R, then $(d) = \cap_{i=1}^n P_i^{(e_i)}$, where $\{P_1, \ldots, P_n\}$ is the complete set of minimal regular prime ideals of R containing d.

PROOF: By Lemma 8.1 $dR_{(P_i)} = P_i^{e_i} R_{(P_i)}, i = 1, \ldots, n$. Therefore $(d) = d(\cap R_{(P_\lambda)}) \cap R = \cap_{i=1}^n (dR_{(P_i)} \cap R) = \cap_{i=1}^n P_i^{(e_i)}$.

Notes

Section 5 The first extension of valuation theory to arbitrary commutative rings was given by Samuel [101]. For a ring T and a subring R of T, he introduced the following three conditions.

(P_1) There exists a prime ideal P of R such that for each ring B, $R \subset B \subseteq T$ and for each prime ideal Q of B, $Q \cap R \neq P$ (cf. condition (1) of Theorem 5.1).

(P_2) $T \setminus R$ is multiplicatively closed.

(P_3) R has the dominated polynomial property with respect to R.

Samuel proved that $(P_1) \Rightarrow (P_3) \Rightarrow (P_2)$; and that none of these implications were reversible. Manis proved (5.1) [73] and Griffin proved (5.5) [43]. Originally Bourbaki [B] called our paravaluation rings "valuation rings," and Griffin called our paravaluations "evaluations," Manis called the rings satisfying (5.1) valuation rings. Other authors used the term valuation ring to mean a ring whose ideals are linearly ordered. In order to standardize the notation, we have adopted the term valuation rings for the rings displayed in (5.1) and paravaluation rings for rings satisfying (5.5). Rings whose ideals are linearly ordered are studied in Chapter V. These rings are called *chained rings* and form a proper subset of the set of valuation rings.

Section 6 There are two competing generalizations of Prüfer domains to rings with zero divisors other than the one presented in Section 6. First is that of an *arithmetical ring*; that is, a ring R for which $I+(J\cap K) = (I+J)\cap(I+K)$ for all ideals I, J, and K of R. The second generalization is that of a *semihereditary ring*; a ring for which each finitely generated ideal is projective. See Bergman [15] and Marot [76] for characterizations of hereditary and semihereditary rings. Griffin [44], who originally gave the definition of Prüfer ring used in this book, shows that all arithmetical and semihereditary rings can be obtained by placing appropriate restrictions on the total quotient rings of Prüfer rings; and, in fact, the set of Prüfer rings properly contains the other two classes of rings. In addition, our definition of Prüfer ring is sufficiently rich enough to capture the multiplicative ideal theory that one would expect a generalization of a Prüfer domain to possess.

The idea of the large quotient ring also comes from [44]. In this book only a few of the equivalent conditions for a ring R to be Prüfer are needed. These are given in (6.2). There are many others; a list of which may be found in [LM]. The proofs of the equivalent conditions for a Prüfer ring are due to Butts and Smith [19] and to Griffin [44]. D. D. Anderson and J. Pascual [8] have developed the theory of Prüfer rings (same definition as ours) in which they characterize these rings solely in terms of their regular ideals; for example, they show that condition (4) of Theorem 6.2 may be replaced by: If $IJ = IK$ where I, J, K are regular ideals of R and I is finitely generated, then $J = K$. Theorem 6.5 is proved by Boisen and Larsen [16].

The first author to discuss Prüfer rings from the point of view of this book was Davis in [25]. His work contained many fruitful ideals that were later expanded on by other authors. Davis gave the following characterization for a special class of Prüfer rings: If R is a ring with few zero divisors, each overring of R is integrally closed if and only if $R_{(M)}$ is a valuation ring for each maximal ideal M of R (cf. Theorem 6.3).

Another point of view is taken in the papers by Eggert [28], and Churchel and Eggert [20]. In these papers the total quotient ring $T(R)$ of R is replaced by the complete ring of quotients $Q(R)$. Then the valuation-Prüfer theory is considered for R with respect to $Q(R)$ (instead of $T(R)$). Some interesting differences occur. An investigation of $Q(R)$, for a Prüfer ring R, is then carried out.

Section 7 Marot was the first to systematically extend the multiplicative theory of ideals from integral domains to commutative rings with zero divisors [75]. In his work he made exceptional use of the following axiom: Each regular ideal is generated by its set of regular elements. He called this property (P). This terminology is still used by many authors, but more recently the term "Marot ring" is being used for rings with property (P). I think this is entirely appropriate and have adopted it for this book. Additively regular rings were implicit in [75], and were named by Gilmer and Huckaba [39]. Theorems 7.1 and 7.4 are in [75], while 7.5 is in [55]. Portelli and Spangher proved (7.6) and (7.10) [94]. Theorem (7.9) is due to Hinkle and Huckaba [51].

Section 8 Krull rings were introduced in [75]. The assumption of the Marot property is crucial in our development of Krull rings. There are many reasons for this. One is that the hypothesis is needed to prove that a divisorial ideal is the intersection of the principal fractional ideals containing it (8.3). A counterexample exists to this result when the Marot hypothesis is deleted; see Example 11 of Section 27. The work on divisors and divisor class groups comes from [G], with suitable modifications. Theorem 8.6 is due to Kennedy [59]. The converse of (8.6) (not presented here) is true and is proved by Matsuda in [80]. The material covered from (8.7) to (8.13) is a reworking of the corresponding material for integral domains that appears in Fossum's book [Fo]. D. D. Anderson, D. F. Anderson and R. Markanda [6] proved (8.16).

Ratliff's work on prime divisors of the integral closure of a principal ideal [98] contains another possible generalization of Krull domains.

The ring theoretic conditions listed in Section 5 for valuation rings, Section 6 for Prüfer rings, and Section 8 for Krull rings are similar to the corresponding conditions for integral domains. However there are serious differences between the ring and domain cases. Some of these have been noted in the text; for example, Theorem 8.16. Others are discussed in the example section.

General references for the material appearing in Chapter II are [75], [94], and Matsuda's survey article [84].

Chapter III

Integral Closure

9. Integral Closure as Intersections

Krull proved in the early 1930s that the integral closure of a domain D is the intersection of the valuation domains between D and its quotient field. This section investigates Krull's theorem in the case that the ring R possesses zero divisors. The first result extends Krull's theorem in terms of paravaluation rings between R and T, where T is not necessarily $T(R)$.

Theorem 9.1 If T is an extension ring of R, then the integral closure R' of R in T is the intersection of the paravaluation rings of T that contain R.

PROOF: If $\{V_\alpha\}$ is the set of paravaluation subrings of T that contain R, then $R' \subseteq \cap V_\alpha$. Choose $a \in T$ and assume that a is not integral over R. It is easy to see that the set I of elements

$x \in T$ such that $a^n x = 0$, for some positive integer n, is an ideal of T. Nilpotent elements of T are integral over R, hence $a \notin I$. Let $\bar{T} = T/I$ and let $\bar{a}$ be the image of a in $\bar{T}$. Clearly $\bar{a}$ is regular in $\bar{T}$. It follows from the fact that a is not integral over R that $\bar{a}$ is not integral over $\bar{R} = (R + I)/I$. Then $\bar{R}[\bar{a}^{-1}]$ is a subring of $\bar{T}[\bar{a}^{-1}]$, and $\bar{a}^{-1}\bar{R}[\bar{a}^{-1}]$ is a proper ideal of $\bar{R}[\bar{a}^{-1}]$. Choose $\bar{P}$ to be a proper prime ideal of $\bar{R}[\bar{a}^{-1}]$ containing $\bar{a}^{-1}$. Let $(\bar{V}, \bar{M})$ be a valuation ring of $\bar{T}[\bar{a}^{-1}]$ such that $\bar{V} \supseteq \bar{R}[\bar{a}^{-1}]$ and $\bar{M} \cap \bar{R} = \bar{P}$, Theorem 5.2. Suppose that $\bar{v}$ is the valuation associated with $\bar{V}$ and G is its value group. Let $\bar{v}_0$ be the restriction of $\bar{v}$ to $\bar{T}$. It is easily seen that $\bar{v}_0$ is a paravaluation on the ring $\bar{T}$. Define $v_0: T \to G \cup \{\infty\}$ via $v_0(t) = \bar{v}_0(\bar{t})$. Then v_0 is a paravaluation on T. If V is its associated ring, then $V = V_\alpha$ for an appropriate α. Note that $a \notin V$, since $v_0(a) < 0$. Therefore $R' = \cap V_\alpha$.

Corollary 9.2 Let R be a ring with total quotient ring $T(R)$. Then R is integrally closed if and only if R is the intersection of its set of paravaluation overrings.

There are examples to show that Theorem 9.1 is not true when "paravaluation" is replaced by "valuation." In fact it is possible to construct such an example where R and T are integral domains; see Chapter VI. The more interesting case is whether or not "paravaluation" can be replaced by "valuation" in the statement of Corollary 9.2. The answer to this is not known. However, the replacement can be made for Marot rings.

Theorem 9.3 Let R be a Marot ring with total quotient ring $T(R)$. Then R is integrally closed if and only if R is the intersection of the valuation rings containing it.

PROOF: Use Theorem 7.7 and Corollary 9.2.

Theorem 9.3 is very important in the rest of this book.

10. Krull Rings II

This section continues the study of Krull rings; concentrating on Noetherian rings, or rings that are closely related to Noetherian rings.

Theorem 10.1 The integral closure of a Noetherian ring is a Krull ring.

PROOF: Let R be a Noetherian ring. First, assume that R is reduced. Let $\{Z_i\}_{i=1}^{g}$ be the set of minimal prime ideals of R. If K_i is the quotient field of R/Z_i, then $T(R)$ is the direct sum of the K_i, and R' is the direct sum of the $(R/Z_i)'$. Each $(R/Z_i)'$, being the integral closure of a Noetherian domain, is a Krull domain. Write $(R/Z_i)' = \cap V_{i_j}$, where $\{V_{i_j}\}_{j\in J}$ is the defining family of discrete rank one valuation domains for $(R/Z_i)'$. Define

$$W_{i_j} = K_1 \oplus \cdots \oplus K_{i-1} \oplus V_{i_j} \oplus K_{i+1} \oplus \cdots \oplus K_g$$

The value group of V_{i_j} is the additive group of integers Z. The mapping

$$w_{i_j} : K_1 \oplus \cdots \oplus K_i \oplus \cdots \oplus K_g \longrightarrow Z \cup \{\infty\}$$

given by $w_{i_j}((a_1, \ldots, a_i, \ldots, a_g)) = v_{i_j}(a_i)$ is a discrete rank one valuation on $T(R)$, and $W_{i_j} = \{x \in T(R) : w_{i_j}(x) \geq 0\}$. It follows that $R' = \cap W_{i_j}$, and for each regular $x \in T(R)$, $w_{i_j}(x) = 0$ for all but a finite number of w_{i_j}. Let M_{i_j} be the maximal ideal of V_{i_j} and define

$$P_{i_j} = K_1 \oplus \cdots \oplus K_{i-1} \oplus M_{i_j} \oplus K_{i+1} \oplus \cdots \oplus K_g$$

Then P_{i_j} is the unique maximal regular ideal of W_{i_j}. If $Q = P_{i_j} \cap R'$, then Q is regular and $R'_{(Q)} = W_{i_j}$. This proves that R' is a Krull ring (when R is assumed to be reduced).

Assume that the nilradical N of R is nonzero. If N' denotes the nilradical of R', then $N' \cap R = N$. Canonically embed R/N into R'/N' and embed R'/N' into $(R/N)'$ via the mapping $x/y + N' \mapsto (x+N)/(y+N)$. Similarly embed $T(R)/N'T(R)$ into $T(R'/N')$. It may be assumed without loss of generality that:

$$R/N \subseteq R'/N' \subseteq (R/N)' \subseteq T(R/N) = T(R'/N') \qquad (10a\ )$$

$$T(R)/N' = T(R)/N'T(R) \subseteq T(R'/N') \qquad (10b\ )$$

By the first part of the proof, there is a family of discrete rank one valuations $\{v_i\}$, with corresponding valuation rings $\{V_i\}$, such that $(R/N)' = \cap V_i$.

Let V be one of the V_i. In view of relation (10b), define a mapping $w: T(R) \to Z \cup \{\infty\}$ by $w(x) = v(x + N')$. Conditions (a), (b), and (c) of the definition of a valuation are satisfied by w. Let $W = \{x \in T(R) : w(x) \geq 0\}$. Then W is a paravaluation overring of the Marot ring R, hence W is a valuation ring, Theorem 7.7. Therefore, for each V_i, a valuation overring W_i of R may be derived. It is clear that the W_i are discrete rank one and that $R' \subseteq \cap W_i$. Assume there is an element $x \in \cap W_i \setminus R'$. The coset $x + N' \notin R'/N'$. Since R'/N' is the integral closure of R/N in $T(R)/N'$, $T(R)/N'$ is a subring of $T(R/N)$, and $(R/N)' \cap T(R)/N' = R'/N'$; it follows that $x + N' \notin (R/N)'$. Therefore $x + N'$ is not in some V_i, hence $x \notin W_i$ for some i. This contradiction proves that $\cap W_i = R'$. That $x \in T(R)$ is a unit in all but finitely many W_i follows from the fact that $x + N'$ is a unit in all but finitely many V_i. Therefore R' is a Krull ring.

If I is an ideal in R, then an element $x \in R$ is *integrally dependent* on I in case x satisfies an equation $x^n + c_1 x^{n-1} + \cdots + c_n = 0$, where $c_i \in I^i$. From the Exercises in [N, p. 34], the set

$I_a = \{x \in R : x$ integrally dependent on $I\}$ is an ideal of R such that $I \subseteq I_a \subseteq \operatorname{Rad} I$.

Lemma 10.2 Let N be the nilradical of R and suppose that b is a regular element in R. Then

(1) $(bR)_a = bR' \cap R$.
(2) $N \subseteq (bR)_a$ and $(bR)_a/N = ((b+N)R/N)_a$.
(3) A necessary and sufficient condition for $R = R'$ is $(bR)_a = bR$, for every regular element $b \in R$.

PROOF: (1): Note that x is in $(bR)_a$ if and only if $x^n + r_1 bx^{n-1} + \cdots + r_{n-1}b^{n-1}x + r_n b^n = 0$, where $r_i \in R$. Thus $x/b \in R'$ if and only if $x \in bR' \cap R$.

(2): Since each element of N is nilpotent, $N \subseteq (bR)_a$. Let $x + N \in (bR)_a/N = (bR' \cap R)/N$, so $x + N = bz + N$ with $z \in R'$, and whence $x + N \in (b+N)(R/N)' \cap R/N = ((b+N)R/N)_a$.

For the other inclusion, choose $x + N \in ((b+N)R/N)_a$. Then $x^n + a_1 x^{n-1} + \cdots + a_n = t \in N$ where $a_i \in (bR)^i$. Let $t^k = 0$. It follows that $x^{nk} + c_1 x^{nk-1} + \cdots + c_{nk} = 0$ where $c_i \in (bR)^i$. Therefore $x + N \in (bR)_a/N$.

(3): Straightforward.

The *approximation theorem for Krull rings* is needed. The proof is modeled closely after the proof for domains [G, p. 540]. Notice the additively regular hypothesis.

Theorem 10.3 Let R be an additively regular Krull ring, let $\{P_\lambda\}_{\lambda \in \Lambda}$ be the set of minimal regular prime ideals of R, and let v_λ be the valuation associated with $R_{(P_\lambda)}$. If $v_1, \ldots, v_n$ is a finite subset of $\{v_\lambda\}_{\lambda \in \Lambda}$ and if $k_1, \ldots, k_n$ are integers, then there exists a regular element $t \in T(R)$ such that $v_i(t) = k_i$ for $i = 1, \ldots, n$, and $v_\lambda(t) \geq 0$ for the other v_λ.

PROOF: It suffices to prove the result under the assumption that at most one of the $k_i \neq 0$. For suppose there exist elements

$t_1, \ldots, t_n \in T(R)$ such that

$$v_i(t_j) = \begin{cases} k_i & \text{if } i = j \\ 0 & \text{if } i \neq j \end{cases}$$

and such that $v_\lambda(t_j) \geq 0$ for the rest of the v_λ. Then $t = t_1 t_2 \cdots t_n$ is the required element.

Therefore we prove the theorem under the assumption that $k_2 = \cdots = k_n = 0$. If $k_1 = 0$, let $t = 1$.

Assume that $k_1 > 0$. Choose $y \in P_1^{(k_1)} = P_1^{k_1} R_{(P_1)} \cap R$, $y \notin P_1^{(k_1+1)} \cup P_2 \cup \cdots \cup P_n$. Pick b to be a regular element in $P_1^{(k_1+1)} \cap P_2 \cap \cdots \cap P_n$. Since R is additively regular, there exists $u \in R$ such that $y + bu = t$ is a regular element. Then $v_1(t) = k_1$, $v_i(t) = 0$, $i = 2, \ldots, n$, and $v_\lambda(t) \geq 0$ for the other v_λ.

Assume that $k_1 < 0$. Choose a regular element $y \in T(R)$ such that $v_1(y) = -k_1$, $v_i(y) = 0$, $i = 2, \ldots, n$, and $v_\lambda(y) \geq 0$ for the rest of the v_λ. Then $v_1(y^{-1}) = k_1$, $v_i(y^{-1}) = 0$, $i = 2, \ldots, n$, and $v_\lambda(y^{-1}) \leq 0$ for the other v_λ. There are only finitely many members of $\Lambda \setminus \{1, \ldots, n\}$; say $\{n+1, \ldots, s\}$, such that $v_i(y^{-1}) = -h_i < 0$. By the earlier case, choose a regular element t' such that $v_i(t') = 0$, $i = 1, \ldots, n$, $v_i(t') = h_i$, $i = n+1, \ldots, s$, and $v_\lambda(t') \geq 0$, for the rest of the v_λ. If $t = y^{-1}t'$, then $v_1(t) = k_1$, $v_i(t) = 0$, $i = 2, \ldots, n$ and $v_\lambda(t) \geq 0$ for the other v_λ.

Theorem 10.4 Let R be an additively regular Krull ring. If P is a minimal regular prime ideal of R such that R/P is a Noetherian ring, then $R/P^{(e)}$ is a Noetherian ring for each positive integer e.

PROOF: Let $\{P_\lambda\}$ be the minimal regular prime ideals of R and let v_λ be the valuation associated with $R_{(P_\lambda)}$. Then $\{v_\lambda\}$ is the defining family of valuations for R. By the approximation theorem (Theorem 10.3), there exists a regular $y \in T(R)$ such that $v_P(y) = -1$ and $v_\lambda(y) \geq 0$ for all $P_\lambda \neq P$. If $x = 1/y$ then $v_P(x) = 1$ and

$v_\lambda(x) \leq 0$ for the rest of the λ. Map $R \to R[x] \to R[x]/xR[x]$ where the first mapping is inclusion and the second is canonical. Call the composite function φ. It is easy to see that φ is surjective and that $\ker \varphi = P$. Therefore $R/P = R[x]/xR[x]$ is a Noetherian ring.

Fix a positive integer e. Map $R \to R[x] \to R[x]/x^e R[x]$, again where the first map is inclusion and the second is canonical. This time the composite map Ψ has kernel $P^{(e)}$, the eth symbolic power of P; but Ψ is not necessarily surjective. We show that $R[x]/x^e R[x]$ is Noetherian by proving that each of its prime ideals are finitely generated [K, p. 5]. Let $\bar{Q} = Q/x^e R[x]$ be a prime ideal of $R[x]/x^e R[x]$, where Q is a prime ideal of $R[x]$ containing $x^e R[x]$. But $x^e \in Q$ implies that $x \in Q$; hence $Q/xR[x]$ is finitely generated, since $R/xR[x]$ is a Noetherian ring. Therefore Q, and hence $\bar{Q}$, is finitely generated.

Note that $R[x]/x^e R[x]$ is a Noetherian ring containing $R/P^{(e)}$ that is finitely generated as an $R/P^{(e)}$-module (the generators are $1, \bar{x}, \ldots, \bar{x}^{e-1}$). By Eakin's theorem [K, p. 54, Ex. 15], $R/P^{(e)}$ is a Noetherian ring.

The preceding result yields an important corollary.

Corollary 10.5 Let R be an additively regular Krull ring. Suppose that for each minimal regular prime ideal P of R, R/P is a Noetherian ring. Then for each regular element $r \in R$, R/rR is a Noetherian ring. Hence every regular ideal of R is finitely generated.

PROOF: Let $(r) = P_1^{(e_1)} \cap \cdots \cap P_n^{(e_n)}$ where $P_1, \ldots, P_n$ are minimal regular prime ideals of R containing (r), Theorem 8.17. By Theorem 10.4, each $R/P_i^{(e_i)}$ is a Noetherian ring. Therefore $R/(r) = R/(\cap P_i^{(e_i)})$ is a Noetherian ring [N, p. 11].

11. Integral Closure of Noetherian Rings I

The purpose of the next two sections is to extend the following well-known theorem of Nagata from integral domains to arbitrary rings.

> $(*)$ [N, Theorems 33.2 and 33.12]. If D is a Noetherian domain of dimension ≤ 2, then D' is a Noetherian domain.

Our first objective is to show that there is no possibility of generalizing $(*)$ by replacing "domain" by "ring." This is done by showing how to construct a Noetherian ring R of any positive dimension such that R' is a non-Noetherian ring. First, a lemma.

Lemma 11.1 Let R be a ring and let $J(R)$ denote its Jacobson radical. If R' is a Noetherian ring, then $J(R) \subseteq Z(R)$ or $N(R) = (0)$.

PROOF: Assume that there exists an element $b \in J(R) \setminus Z(R)$. Then b is a regular element of R'. If $x \in N(R')$, then x/b is a nilpotent element of $T(R) = T(R')$; and since R' is integrally closed, $x/b \in N(R')$. Consequently $N(R') = bN(R')$. By Nakayama's lemma [K, p. 51], $N(R') = 0$. Hence $N(R) = N(R') \cap R = (0)$.

For the example let D be an n-dimensional local Noetherian domain, $n \geq 2$. Assume that P is a height one prime ideal of depth $n-1$, and assume that A is a P-primary ideal distinct from P. If $R = D/A$, then the preceding lemma implies that R' is a non-Noetherian ring of dimension $n - 1$.

Hence there are one- and two-dimensional Noetherian rings with the property that their integral closures are non-Noetherian.

We concentrate on two generalizations of $(*)$. In this section the following is proved.

(**) If R is a Noetherian ring of dimension ≤ 2, then each regular ideal of R' is finitely generated.

The second generalization of (*), necessary and sufficient conditions for R' to be a Noetherian ring, is delayed until Section 12.

The next theorem and its corollary gives the one-dimensional version of (**).

Theorem 11.2 Let S be an overring of a one-dimensional Noetherian ring R. If the nilpotent elements of $T(R)$ are in S, then every regular ideal of S is finitely generated.

PROOF: There is a one-to-one correspondence between $\{Z_i\}_{i=1}^{g}$, the minimal prime ideals in R; and $\{Z_i'\}_{i=1}^{g}$, the minimal prime ideals in S. In this correspondence $Z_i' \cap R = Z_i$. Therefore

$$R/Z_i \subseteq S/Z_i' \subseteq T(R/Z_i) \quad (i = 1, \ldots, g)$$

By the Krull–Akizuki theorem [N, p. 115], each S/Z_i' is a Noetherian domain, and hence $S/(\cap Z_i')$ is a Noetherian ring [N, p. 11]. Let I be a regular ideal in S and let π be the canonical homomorphism of S onto $S/(\cap Z_i')$. Choose $a_1, a_2, \ldots, a_n$ in I, a_1 regular, such that $\{\pi(a_i)\}_{i=1}^{n}$ generates $\pi(I)$. If $z \in I$, then $\pi(z) = \sum_{i=1}^{n} \pi(c_i a_i)$, where $c_i \in S$. By hypothesis, $z - \sum c_i a_i \in \cap Z_i' = N(S) \subseteq a_1 S$. Thus I is finitely generated.

Corollary 11.3 Let R be a one-dimensional Noetherian ring; then every regular ideal in R' is finitely generated.

Section 12 contains a complete generalization of the Krull–Akizuki theorem for rings with zero divisors.

Theorem 11.4 If R is a Noetherian ring and P a prime ideal of R, then there are only finitely many prime ideals of R' that lie over P.

PROOF: First, assume that R is a reduced ring with minimal prime ideals $\{Z_i\}_{i=1}^{g}$. Then $R' = D_1 \oplus \cdots \oplus D_g$ where $D_i = (R/Z_i)' \cong R'/Z_i'$. If $Q' \in \operatorname{Spec} R'$ such that $Q' \cap R = P$, then $Q' = D_1 \oplus \cdots \oplus q \oplus \cdots \oplus D_g$ where q is a prime ideal in $(R/Z_i)' \cong R'/Z_i'$. Write $q \cong Q/Z_i'$ with Q a prime ideal in R'. It follows that $Q/Z_i' \cap (R + Z_i')/Z_i' = (P + Z_i')/Z_i'$. By the domain case there are only finitely many prime ideals of $(R/Z_i)'$ lying over P/Z_i, and there are only finitely many Z_i; so the result holds.

Next, assume that the nilradical N of R is nonzero and let N' be the nilradical of R'. If P is a prime ideal of R, let $\bar{P}$ denote the prime ideal P/N of R/N. From the reduced case, there are only finitely many prime ideals $\{\bar{Q}_i\}_{i=1}^{n}$ of $(R/N)'$ lying over $\bar{P}$. By (10a), $\{\bar{Q}_i \cap R'/N'\}$ is the complete set of prime ideals of R'/N' lying over $\bar{P}$. If $\pi : R' \to R'/N'$ is the canonical homomorphism, let $Q_i = \pi^{-1}(\bar{Q}_i \cap R'/N')$. Then $\{Q_i\}_{i=1}^{n}$ is the set of prime ideals of R' lying over P.

Theorem 11.5 Let R be a Noetherian ring. For each P' of R', $T(R'/P')$ is a finite field extension of $T(R/P' \cap R))$.

PROOF: Let $N' = N(R'), N = N(R)$, and $P = P' \cap R$. If $Z_1, \ldots, Z_n$ are the minimal prime ideals of R, then $(R/N)' = (R/Z_1)' \oplus \cdots \oplus (R/Z_n)'$, Lemma 8.14. Recall that

$$R/N \subseteq R'/N' \subseteq (R/N)' \subseteq T(R/N)$$

Let Q be a prime ideal of $(R/N)'$ such that $Q \cap R'/N' = P'/N'$. Write

$$Q = (R/Z_1)' \oplus \cdots \oplus (R/Z_{i-1})' \oplus Q_i \oplus (R/Z_{i+1})' \oplus \cdots \oplus (R/Z_n)'$$

where Q_i is a prime ideal of the Krull domain $(R/Z_i)'$. Thus $Q \cap (R/N) = P'/N' \cap (R/N) = P/N$. Consequently $Q_i \cap (R/Z_i) =$

P/Z_i. By [N, p. 118] $T((R/Z_i)'/Q_i)$ is a finite $T((R/Z_i)/(P/Z_i))$-module; that is, $T((R/N)/Q)$ is a finite $T(R/P)$-module. Finally

$$R/P \subseteq R'/P' = (R'/N')/(P'/N') \subseteq (R/N)'/Q$$

so $T(R'/P')$ is also a finite $T(R/P)$-module.

Theorems 10.1, 11.4, and 11.5 give a complete generalization of [N, p. 118]. We end this section with the two-dimensional version of (**).

Theorem 11.6 If R is a two-dimensional Noetherian ring, then every regular ideal of R' is finitely generated.

PROOF: Since R' is an additively regular Krull ring, it suffices to show that for each minimal regular prime ideal P' of R', that R'/P' is Noetherian (Corollary 10.5). Let $P = P' \cap R$; then $\dim R/P \leq 1$, R'/P' is integral over R/P, and $T(R'/P')$ is a finite $T(R/P)$-module (Theorem 11.5). By the Krull–Akizuki theorem [N, p. 115], R'/P' is a Noetherian ring.

The results developed in Sections 10 and 11, culminating with Theorem 11.6, yield a shorter and simpler proof of (*) (even for integral domains) than the proof which appears in [N].

12. Integral Closure of Noetherian Rings II

This section is devoted to the further study of Nagata's theorem which was denoted by (*) in Section 11. We look for necessary and sufficient conditions for 1- and 2-dimensional Noetherian rings to have Noetherian integral closures. In view of Corollary 11.3 and Theorem 11.6, only the prime ideals consisting entirely of zero divisors need be studied.

Let S be an arbitrary subring of $T(R)$ containing R. Assume that R is a Noetherian ring. If

$$(0) = Q_1 \cap \cdots \cap Q_t \quad \text{where } Q_i \text{ is } P_i\text{-primary} \tag{12a}$$

is an irredundant primary representation of (0), and if $P_i T(R) \cap S = P_i'$ and $Q_i T(R) \cap S = Q_i'$, then

$$(0) = Q_1' \cap \cdots \cap Q_t' \quad \text{where } Q_i' \text{ is } P_i'\text{-primary} \tag{12b}$$

is an irredundant primary representation of (0) in S. Note that (12b) always exists even though S may be a non-Noetherian ring.

Assume that R is a Noetherian ring and that $\dim T(R) = 0$. If (12a) is an irredundant primary representation of (0) in R, then $T(R)$ is isomorphic to a direct sum of $T(R)/Q_i T(R) \cong T(R/Q_i)$. Hence

$$T(R) = T(R/Q_1) \oplus \cdots \oplus T(R/Q_t) \tag{12c}$$

Assume, in addition, that (12b) is an irredundant primary decomposition of (0) in R'. Then $R \subseteq R/Q_1 \oplus \cdots \oplus R/Q_t \subseteq (R/Q_1)' \oplus \cdots \oplus (R/Q_t)'$. Arguing as in Lemma 8.14, we see that

$$R' = (R/Q_1)' \oplus \cdots \oplus (R/Q_t)' \tag{12d}$$

Lemma 12.1 Let R be a Noetherian ring and let S be a ring between R and $T(R)$.

(1) If M' is a prime ideal of S consisting of zero divisors, then $R_{R \cap M'} = S_{M'} = T(R)_{M'T(R)}$. In particular $S_{M'}$ is a Noetherian ring.

(2) If M' is a maximal ideal of S consisting entirely of zero divisors such that $M' \cap R$ is a maximal ideal of R, then

M' is a finitely generated ideal. Moreover, if $\dim R \leq 2$, then every maximal ideal of R' is finitely generated.

PROOF: (1): Use [ZSI, p. 231].

(2): Let $M = M' \cap R$ and let $B = MS$. If P' is a prime ideal in S containing B, then $P' \cap R \supseteq B \cap R = M$. Clearly $P' \cap R = M$, $P' \subseteq Z(S)$, and consequently $P' = M'$. Since $MT(R) = M'T(R)$ and $S_{M'} = T(R)_{M'T(R)}$, $MS_{M'} = M'S_{M'}$—that is, $BS_{M'} = M'S_{M'}$. But M' is the only prime ideal of S containing B, so $M' = B$. This proves that M' is finitely generated. The second conclusion follows from the first and Theorem 11.6.

The equalities in (12c), (12d), and part 1 of Lemma 12.1 are actually isomorphisms. Since nothing is lost by writing equality in these situations, we will continue to do so.

Theorem 12.2 Let R be a Noetherian ring, then the following are equivalent:

(1) $N(R')$ is a finitely generated ideal.

(2) If M' is a regular maximal ideal of R', then $R'_{M'}$ is a reduced ring.

(3) If M is a regular maximal ideal of R, then R_M is a reduced ring.

(4) If P is an associated prime divisor of (0) and if P is contained in a regular maximal ideal of R, then P is a minimal prime divisor of (0) and the P-primary component of (0) is P.

Furthermore, the above conditions are necessary for the ring R' to be Noetherian.

PROOF: (1) $\Rightarrow$ (2), (3): Let M' be a regular maximal ideal of R', $M = R \cap M'$, and $S = R \setminus M$. Then R'_S is the integral closure of R_M in $T(R)_S$ [G, p. 91], and $N(R'_S) = N(R')R'_S$ is finitely generated. Let b be a regular element in M and let $b/1$ be the image of b under the natural homomorphism $R \to R_M$. Then $b/1$ is in the Jacobson radical of R'_S, and is also a unit in $T(R)_S$. Since $(b/1)N(T(R)_S) = N(T(R)_S)$, we have $(b/1)N(R'_S) = N(R'_S)$. By Nakayama's lemma, $N(R'_S) = (0)$. Thus R'_S and R_M are reduced rings, and so is $(R'_S)_{M'R'_S} = R'_{M'}$.

(2) $\Rightarrow$ (1): There are only a finite number of maximal ideals of R' such that $N(R')R'_{M'}$ is nonzero—namely, the associated prime ideals of (0) that happen to be maximal ideals of R'. Let $M'_1, \ldots, M'_r$ be these maximal ideals. Since each $R'_{M'_i}$ is a Noetherian ring (Lemma 12.1), a finite number of elements $a_{i1}, \ldots, a_{iu_i}$ in $N(R')$ can be chosen so that their images in $R'_{M'_i}$ generate $N(R')R'_{M'_i}$. Let B be the ideal in R' generated by the set $\{a_{ij}\}$ $i = 1, \ldots, r$; $j = 1, \ldots, u_i$. Then B is a finitely generated ideal of R' that agrees locally with $N(R')$. Thus $B = N(R')$.

(3) $\Rightarrow$ (2): With the same notation as (1) $\Rightarrow$ (2), (3), R_M is reduced implies R'_S is reduced, whence $R'_{M'}$ is reduced.

(3) $\Leftrightarrow$ (4): This follows directly from the behavior of irredundant primary decompositions under localizations.

Finally, if R' is a Noetherian ring, then condition (1) (equivalently: (2), (3), (4)) holds.

The next result characterizes those one- and two-dimensional Noetherian local rings whose integral closures are Noetherian.

Theorem 12.3 Let (R, M) be a local Noetherian ring of dimension ≤ 2, where M is a regular ideal. The following conditions are equivalent:

(1) R' is a Noetherian ring.
(2) R' is a locally Noetherian ring.
(3) If M' is a maximal ideal of R', then $R'_{M'}$ is a reduced ring.
(4) If M' is a maximal ideal of R', then $R'_{M'}$ is an integral domain.

(5) R' (equivalently R) is a reduced ring.
(6) $N(R')$ is a finitely generated ideal of R'.
(7) If P is an associated prime ideal of (0) in R, then P is a minimal prime divisor of (0) and the P-primary component of (0) is P.
(8) R' is a finite direct sum of Noetherian integrally closed domains.

PROOF: Every associated prime ideal of (0) in R is contained in the regular maximal ideal M of R, and R' has only finitely many maximal ideals, all of which are regular, Theorem 11.4.

The implications (1) $\Rightarrow$ (2), (8) $\Rightarrow$ (1) and (4) $\Rightarrow$ (3) are obvious. Theorem 12.2 gives the equivalence of (3), (5), (6), and (7).

(2) $\Rightarrow$ (1): The locally Noetherian ring R' has only finitely many maximal ideals. Hence it must be a Noetherian ring.

(1) $\Rightarrow$ (5): The Jacobson radical of R is a regular ideal. By Lemma 11.1, $N(R) = 0$.

(5) $\Rightarrow$ (8): Use Lemma 8.14 and the fact that the integral closure of a one- or two-dimensional Noetherian integral domain is Noetherian.

(8) $\Rightarrow$ (4): This implication follows directly from properties of finite direct sums of integral domains.

Let S be the complete direct sum of a family of rings $\{S_\alpha\}$ and let $\pi_\alpha : S \to S_\alpha$ be the canonical epimorphism. A subring R of S is called a *subdirect sum* of $\{S_\alpha\}$ in case $\pi_\alpha(R) = S_\alpha$, for each α.

The global characterizations follow.

Theorem 12.4 Let R be a Noetherian ring of dimension ≤ 2 and assume that $\dim T(R) \leq 1$. The following conditions are equivalent:

(1) R' is a Noetherian ring.
(2) R' is a locally Noetherian ring.

(3) If M' is a regular maximal ideal of R', then $R'_{M'}$ is a reduced ring.
(4) If M' is a regular maximal ideal of R', then $R'_{M'}$ is an integral domain.
(5) If M is a regular maximal ideal of R, then R_M is a reduced ring.
(6) $N(R')$ is a finitely generated ideal of R'.
(7) If P is an associated prime ideal of (0) and if P is contained in a regular maximal ideal of R, then P is a minimal prime divisor of (0) and the P-primary component of (0) is P.
(8) R' is a subdirect sum of a finite number of Noetherian integral domains and a Noetherian total quotient ring.

If $\dim T(R) = 0$ then (1) through (8) are equivalent to:

(8′) R' is a finite direct sum of Noetherian integrally closed domains and a Noetherian total quotient ring.

PROOF: By Theorem 12.2, (3), (5), (6), and (7) are equivalent.
(1) $\Rightarrow$ (2): Trivial.
(4) $\Rightarrow$ (7): Theorem 12.2.
(5) $\Rightarrow$ (4): Let M' be a regular maximal ideal of R', $M' \cap R = M$, and $S = R \setminus M$. Then R'_S is the integral closure of R_M in $T(R)_S$.

We show that $T(R)_S = T(R_M)$. Since $T(R)$ is a Noetherian total quotient ring of dimension ≤ 1, every prime ideal in $T(R)$ is an associated prime of (0). From the fact that S is a multiplicatively closed subset of $T(R)$, it follows that every prime ideal in $T(R)_S$ is an associated prime of $(0)_S$, hence $T(R)_S$ is a total quotient ring containing R_S. Any regular element in R_S remains regular in $T(R)_S$, and is therefore a unit in $T(R)_S$. Therefore $R_S \subseteq T(R_S) \subseteq T(R)_S$. A typical element of $T(R)_S$ is of the form $(a/r)/s$, where $a/r \in T(R)$ and $s \in S$. But r is regular in R, hence $r/1$ is regular in R_S, and so it is a unit in $T(R)_S$.

Consequently $(a/r)/s = (a/s)(r/1)^{-1} \in T(R_M)$. This proves that $T(R_M) = T(R)_S$.

Since $\dim T(R_M) = 0$, R'_S is a direct sum of integral domains, Lemma 8.14. Therefore the reduced ring $R'_{M'} = (R'_S)_{M'R'_S}$ is a domain, since $M'R'_S$ contains a unique minimal prime ideal of R'.

(2) $\Rightarrow$ (3): For a regular maximal ideal M' of R', again let $M = M' \cap R$ and $S = R \setminus M$. Then R'_S is a semilocal ring which is locally Noetherian, hence R'_S is Noetherian. But, as in (5) $\Rightarrow$ (4), R'_S is the integral closure of R_M in $T(R_M) = T(R)_S$. By Theorem 12.3, $R'_{M'} = (R'_S)_{M'R_S}$ is a reduced ring.

(7) $\Rightarrow$ (1): By the hypothesis, the irredundant primary decomposition of (0) in R' can be written as

$$(0) = P'_1 \cap \cdots \cap P'_s \cap Q'_{s+1} \cap \cdots \cap Q'_t \tag{12e}$$

where P'_i $(1 \leq i \leq s)$ are the associated prime ideals of (0) contained in regular ideals and Q'_i $(s+1 \leq i \leq t)$ are primary components of (0) not contained in regular ideals of R'.

Consider a proper prime ideal $\bar{P}'$ of R'/P'_i for some $i \leq s$. Let P' be the corresponding prime ideal in R'. If P' is a regular ideal, then it is finitely generated. If $P' \subseteq Z(R')$, then since $P' \supset P'_i$ and $\dim T(R) \leq 1$, P' is a maximal ideal of R'. By Lemma 12.1, P' is finitely generated. In either case $\bar{P}'$ is finitely generated. Cohen's theorem [K, p. 5] implies that R'/P'_i is a Noetherian ring.

Let $C = Q'_{s+1} \cap \cdots \cap Q'_t$ and let $C^e = CT(R)$. Then R'/C is a total quotient ring. (For suppose that M' is a maximal ideal of R' that contains C; then M' contains some Q'_i, and hence $M' \subseteq Z(R')$. Thus M' is an associated prime of (0). Therefore the image of M' in R'/C consists entirely of zero divisors.) It follows that $T(R'/C) = R'/C \subseteq T(R')/C^e \subseteq T(R'/C)$. Hence $R'/C = T(R)/C^e$ is a Noetherian ring. Since $P'_1 \cap \cdots \cap P'_s \cap C = (0)$, R' is a Noetherian ring [N, p. 11].

(7) $\Rightarrow$ (8): Consider the irredundant primary decomposition of (0) given by equation (12e). If $C = Q'_{s+1} \cap \cdots \cap Q'_t$, then R' is a subdirect sum of $R'/P'_1 \oplus \cdots \oplus R'/P'_s \oplus R'/C$.

(8) $\Rightarrow$ (1): We must prove that if S is a subdirect sum of $S_1 \oplus \cdots \oplus S_n$, where S_j is a Noetherian ring, then S is a Noetherian ring. Let $\alpha : S \to S_1 \oplus \cdots \oplus S_n$ be the embedding monomorphism and let $\pi_j : S_1 \oplus \cdots \oplus S_n \to S_j$ be the canonical projections. Then $S/\ker(\pi_j\alpha) \cong S_j$, for all j. But $\cap_{j=1}^n \ker(\pi_j\alpha) = (0)$, so S is Noetherian.

Clearly (8′) $\Rightarrow$ (1). Let $\dim T(R) = 0$ and assume (7). Equations (12d) and (12e) imply that $R' = (R/P_1)' \oplus \cdots \oplus (R/P_s)' \oplus R'/C$, where $C = Q'_{s+1} \cap \cdots \cap Q'_t$.

It is not known whether Theorem 12.4 holds when the hypothesis that $\dim T(R) \leq 1$ is deleted.

If R is a Noetherian ring, let T be the set of all elements $x \in T(R)$ such that $(R :_R x)$ contains a finite product of maximal ideals of R. Call T the *global transform* of R. The classical Krull–Akizuki theorem says (among other things) that if R is a one-dimensional Noetherian domain, then each overring S of R is Noetherian. This is equivalent to showing that if x is a nonzero element in R, then S/xS is a finite R-module [N, p. 115]. We give two generalizations of the Krull–Akizuki theorem. The first is given by deleting the domain and dimension hypothesis on R and replacing the quotient field of R by the global transform T. This result is then used to give the second generalization; namely, necessary and sufficient conditions for the classical result to hold for rings with zero divisors.

Theorem 12.5 If R is a Noetherian ring, T is the global transform of R, and S is a ring between R and T, then S/xS is a finite R-module for each regular $x \in R$.

PROOF: Fix a regular element x in R. For $a \in S$ let $J =$

$(R :_R a)$. Then J contains a finite product of maximal ideals of R. If $M_1, \ldots, M_t$ are maximal ideals of R such that $(M_1 M_2 \cdots M_t)^r \subseteq J$, then each maximal ideal containing J is some M_i. Hence R/J is an Artinian ring [ZSI, p. 203]. Choose a positive integer k such that $(x^k) + J = (x^{k+1}) + J$ in R/J. Write $x^k = rx^{k+1} + j$ for some $r \in R$ and $j \in J$. Then $ax^k = arx^{k+1} + aj$ which implies that $a \in xS + Rx^{-k}$.

Since R is a Noetherian ring, $(xS \cap R)/xR$ is a finite R-module. Say that $xa_1, \ldots, xa_w \in xS \cap R$ are picked so that the $xa_i + xR$ generate $(xS \cap R)/xR$. Let I be a finite product of maximal ideals such that $Ia_i \subseteq R$ $(i = 1, \ldots, w)$. Then $(xS \cap R)/xR$ becomes a finite R/I-module, and since R/I is an Artinian ring, $(xS \cap R)/xR$ has finite length as an R/I-module [K, p. 59]. Therefore the descending sequence of ideals $\{I_m\} = \{(x^m S \cap R, xR)\}$ stabilizes for some positive integer n.

Suppose that there is some $a \in S \setminus Rx^{-n} + xS$. By the first paragraph, choose a minimal m, m necessarily $> n$, such that $a \in Rx^{-m} + xS$. Write $a = rx^{-m} + xa'$ where $r \in R$ and $a' \in S$. It follows that $x^m(a - xa') = r \in I_m = I_{m+1}$. Hence $x^m(a - xa') = x^{m+1}a'' + xr'$ with $r' \in R$ and $a'' \in S$. Solving for a yields, $a \in Rx^{-(m-1)} + xS$, a contradiction. Therefore $S/xS \subseteq Rx^{-n}/(Rx^{-n} \cap xS)$ and hence S/xS is a finitely generated R-module.

Theorem 12.6 Let R be a one-dimensional Noetherian ring and let S be a subring of $T(R)$ containing R. Then $\dim S \leq 1$, every regular ideal I of S is finitely generated, and S/I is a finite $R/(I \cap R)$-module. The ring S is Noetherian if and only if its minimal prime ideals are finitely generated. Moreover, conditions (1) through (8) of Theorem 12.4 are equivalent to:

(9) Every subring of $T(R)$ containing R is Noetherian.

PROOF: Let $R \subseteq S \subseteq T(R)$ and let P be a minimal prime ideal of S. Then

$$R/(P \cap R) \subseteq S/P \subseteq T(R)/PT(R) \subseteq T(R/(P \cap R))$$

By the Krull–Akizuki theorem, S/P is a Noetherian domain of dimension ≤ 1, hence $\dim S \leq 1$.

Let T be the global transform of R. If $x = a/b \in T(R)$ with b regular, then every associated prime of (b) is a maximal ideal of R. Let $\{M_1, \ldots, M_t\}$ be these associated primes of (b). Since $(b) \subseteq (R :_R x)$, there exists a positive integer n such that $(M_1 M_2 \ldots M_t)^n \subseteq (R :_R x)$. This proves that $T = T(R)$.

Let I be a regular ideal of S and choose a regular element $x \in I \cap R$. By Theorem 12.5, S/xS is a finite R-module, and is therefore a Noetherian ring. Since $I \supseteq xS$, I is finitely generated and S/I is a finite $R/(I \cap R)$-module.

Assume that the minimal prime ideals of R are finitely generated. We show that S is a Noetherian ring. Since $\dim S \leq 1$, it suffices to prove that the height one maximal ideals of S are finitely generated. Since the regular maximal ideals of R are known to be finitely generated, assume that M' is a height one prime ideal of S consisting of zero divisors. Then $M' \cap R$ is a height one maximal ideal of R, so M' is finitely generated, Lemma 12.1.

The proof that (7) $\Rightarrow$ (9) is similar to the proof that (7) $\Rightarrow$ (1) of Theorem 12.4. The equivalence of (1) through (9) is finished by noting that if every subring of $T(R)$ containing R is Noetherian, then R' is Noetherian.

13. When Polynomial Rings Are Integrally Closed

As the title of this section indicates, we are interested in determining when integral closure is preserved under adjunction of an indeterminate. It is well known (and easy) that if R is an integrally closed domain, then so is the polynomial ring $R[X]$. At the other extreme suppose that R is an integrally closed ring with nonzero nilradical. Then $R[X]$ is never integrally closed; for if a is a nonzero nilpotent element in R, then a/X is integral over, but does not belong to, $R[X]$. Therefore two necessary conditions for $R[X]$ to be integrally closed are that R is reduced and that R is integrally closed. To obtain the results on the integral closedness of $R[X]$, several general theorems about integrally closed rings must be established.

Let M be a maximal ideal of R. Define φ_M to be the homomorphism $T(R) \to T(R_M)$ which is the unique extension of the canonical homomorphism $R \to R_M$.

Lemma 13.1 Assume that $\mathcal{P}$ is a prime ideal of $R[X]$, $\mathcal{P} \cap R = P$, and M is some maximal ideal in R containing P. Then $R_P[X]_{\mathcal{P}R_P[X]} = R_M[X]_{\mathcal{P}R_M[X]} = R[X]_{\mathcal{P}}$.

PROOF: It is easy to see that $R_P[X] = R[X]_{R \setminus P}$ and $R_M[X] = R[X]_{R \setminus M}$. Since $R[X] \setminus \mathcal{P} \supseteq R \setminus P \supseteq R \setminus M$, the result follows.

Lemma 13.2 Let R be a ring. If R_M is integrally closed for each M in $\operatorname{Max} R$, then R is integrally closed.

PROOF: Assume that $a \in T(R)$ and $\varphi_M(a) \in R_M$ for each $M \in \operatorname{Max} R$. We claim that $a \in R$. Suppose that $a = c/b \notin R$. Then there exists a maximal ideal M of R such that $(b :_R c) \subseteq M$. Thus $(b :_R c)R_M = (\varphi_M(b) :_{R_M} \varphi_M(c)) \subseteq MR_M$. We conclude that $\varphi_M(c)/\varphi_M(b) = \varphi_M(a) \notin R_M$, a contradiction.

If $a \in T(R)$ and a is integral over R, then $\varphi_M(a)$ is integral over R_M, and hence in R_M, for each M. By the first paragraph, $a \in R$.

Theorem 13.3 Let R be a reduced ring such that for each $M \in \operatorname{Max} R$, R_M is an integrally closed domain. Then the polynomial ring $R[X]$ is integrally closed.

PROOF: Let $\mathcal{M}$ be a maximal ideal of $R[X]$. It suffices to prove that $R[X]_{\mathcal{M}}$ is integrally closed. By Lemma 13.1, $R[X]_{\mathcal{M}} = R_M[X]_{\mathcal{M}R_M[X]}$, where $M \in \operatorname{Max} R$.

Theorem 13.4 Let R be an integrally closed ring. Then $R[X]$ is integrally closed if and only if $T(R)[X]$ is integrally closed.

PROOF: This follows directly from the fact that $R[X]$ is integrally closed in $T(R)[X]$, [G, p. 96].

We briefly detour into the area of homological algebra. Let R be a ring and M an R-module. For a set I, let $M^I = \prod_{i \in I} M$ and define $\sigma(M) : R^I \otimes M \to M^I$ by $\sigma(M)(\{r_i\} \otimes x) = \{r_i x\}$. An R-module M is *finitely presented* if there exist finitely generated free modules F and K such that $K \to F \to M \to 0$ is exact.

Lemma 13.5 Let M be an R-module; then:

(1) M is finitely generated if and only if $\sigma(M)$ is surjective, for each set I.

(2) M is finitely presented if and only if $\sigma(M)$ is bijective, for each set I.

PROOF: Let

$$E \longrightarrow F \xrightarrow{u} M \longrightarrow O \tag{13a}$$

be exact, where E and F are R-modules and F is free. Then

$$\begin{array}{ccccccc} R^I \otimes E & \longrightarrow & R^I \otimes F & \xrightarrow{1 \otimes u} & R^I \otimes M & \longrightarrow & 0 \\ \big\downarrow{\scriptstyle \sigma(E)} & & \big\downarrow{\scriptstyle \sigma(F)} & & \big\downarrow{\scriptstyle \sigma(M)} & & \\ E^I & \longrightarrow & F^I & \longrightarrow & M^I & \longrightarrow & 0 \end{array} \tag{13b}$$

is a commutative diagram with exact rows.

(1) ($\Leftarrow$): Let $I = M$, then $\sigma(M) : R^M \otimes M \to M^M$. Since $\sigma(M)$ is surjective, there exists elements $\{f_{1m}\}_{m\in M}, \dots, \{f_{nm}\}_{m\in M} \in R^M$ and $x_1, \dots, x_n \in M$ such that $\sigma(M)(\sum_{i=1}^n \{f_{im}\} \otimes x_i) = 1_M$, the identity map on M. Thus $\{\sum_{i=1}^n f_{im}x_i\}_m = 1_M$; hence for $y \in M$, $y = \sum_{i=1}^n f_{iy}x_i$, that is, M is generated by $\{x_1, \dots, x_n\}$.

(1) ($\Rightarrow$): If F is a finitely generated free R-module, then it is easily seen that $\sigma(F)$ is a bijective map and it follows from (13b) that $\sigma(M)$ is surjective.

(2) ($\Rightarrow$): If M is finitely presented, then E and F can be taken as finitely generated free R-modules. Then $\sigma(E)$ and $\sigma(F)$ are bijections implying that $\sigma(M)$ is a bijection.

(2) ($\Leftarrow$): In (13a), let $E = \ker u$ and let F be a finitely generated free R-module (M is finitely generated by Part (1)). Since $\sigma(M)$ and $\sigma(F)$ are bijective, $\sigma(E)$ is surjective [B, p. 8]. By (1), E is finitely generated.

Lemma 13.6 Let $c : R \to T$ be an injective ring homomorphism. If M is a finitely generated flat R-module such that $T \otimes_R M$ is projective as a T-module, then M is a projective R-module.

PROOF: Let I be a set and consider the commutative diagram

$$\begin{array}{ccccccc}
R^I \otimes_R M & \xrightarrow{\alpha} & T^I \otimes_R M & & \xrightarrow{\beta} & & T^I \otimes_T (T \otimes_R M) \\
\downarrow{\scriptstyle \sigma(M)} & & & & & & \downarrow{\scriptstyle \sigma(T\otimes_R M)} \\
M^I & \longrightarrow & R \otimes_R M & \longrightarrow & T \otimes_R M^I & \longrightarrow & T \otimes_R M
\end{array}$$

where $\alpha(\{r_i\} \otimes m) = \{c(r_i)\} \otimes m$, $\beta(\{t_i\} \otimes m) = \{t_i\} \otimes 1 \otimes m$, and the unmarked homomorphisms are canonical. Clearly β is injective and since M is an R-flat module, α is injective. Since $T \otimes_R M$ is a finitely generated projective T-module, it is finitely presented [B, p. 20]. Lemma 13.5 implies that $\sigma(T \otimes_R M)$ is bijective. An easy calculation proves that $\sigma(M)$ is a bijective map.

Hence M is a finitely presented flat R-module, which implies that it is projective [B, p. 45].

The next theorem could have gone in Section 4. But because of its importance to integral closure, we delayed it to here.

Theorem 13.7 Let R be a reduced ring and assume that $\operatorname{Min} R$ is compact. If R_M is an integral domain for each $M \in \operatorname{Max} R$, then $T(R)$ is a von Neumann regular ring.

PROOF: We may assume that $R = T(R)$, Theorem 4.1. For each $x \in R$ and each $M \in \operatorname{Max} R$, xR_M is an R_M-flat module; hence (x) is R-flat [B, p. 92]. Furthermore $Q(R)$ is a flat R-module, Theorem 4.3. Thus $Q(R) \otimes_R (x) \cong xQ(R)$. We already know that $Q(R)$ is a von Neumann regular ring. Consequently $xQ(R)$ is a direct summand of $Q(R)$, hence $xQ(R)$ is a projective $Q(R)$-module. By Lemma 13.6 (x) is a projective R-module. The exact sequence

$$0 \longrightarrow \operatorname{Ann}(x) \xrightarrow{\ i\ } R \xrightarrow{\ g\ } (x) \longrightarrow (0)$$

splits, where i is the inclusion map and g is multiplication by x. Define $h : (x) \to R$ to be the splitting map and suppose that $h(x) = y$. It is easy to see that $\operatorname{Ann}(x) + (y) = R$. In fact this sum is direct; for if $z \in \operatorname{Ann}(x) \cap (y)$, then $z = wy = wh(x) = h(wx)$ and $0 = g(z) = wx$, implying that $0 = h(wx) = z$.

To complete the proof it is sufficient to prove that if $x \in R$, then there exists $b \in R$ such that $bx = 0$ and $b + x$ is a unit of R, Corollary 3.3 (remember $R = T(R)$). Fix $x \in R$ and choose $b \in \operatorname{Ann}(x)$ such that $1 = b + cy$ for some c in R. Clearly $bx = 0$. If $a(b + x) = 0$, then $ab = -ax \in (x) \cap \operatorname{Ann}(x) = (0)$. From $a = ab + acy$ it follows that $a(1 - cy) = 0$. Therefore $a \in (y) \cap \operatorname{Ann}(x) = (0)$, whence $b + x$ is a unit in R.

Theorem 13.8 Let R be a ring whose total quotient ring is von Neumann regular. Then R is integrally closed if and only if for each $M \in \operatorname{Max} R$, R_M is an integrally closed domain.

PROOF: ($\Leftarrow$): Lemma 13.2

($\rightarrow$): Let $S = R \setminus M$, then R_M is integrally closed in $T(R)_S$. Let A be the kernel of the natural homomorphism $R \to R_M$. If $b \in AT(R) \cap R$, write $b = \sum a_i c_i$ where $a_i \in A$ and $c_i \in T(R)$. Choose $s_i \in \operatorname{Ann}(a_i) \cap S$ for each i, and let s be the product of the s_i. Then $sb = 0$, which implies that $b \in A$; that is, $A = AT(R) \cap R$. We may assume that

$$R/A \subseteq T(R)/AT(R) \subseteq T(R/A)$$

Since the homomorphic image of a von Neumann regular ring is von Neumann regular, $T(R)/AT(R) = T(R/A) = T(R)_S$. Next, $R/A \subseteq R_M \subseteq T(R)_S$, so $T(R)_S$ is the total quotient ring of R_M. This proves that R_M is integrally closed.

For the second part, assume that R_M is not an integral domain. Then $T(R_M)$ is a von Neumann regular ring which is not a field. Choose a nonzero nonunit idempotent element $e \in T(R_M)$. Since R_M is integrally closed, both e and $1 - e$ belong to R_M; in fact, to the unique maximal ideal of R_M. Of course this is impossible.

The main results of this section concern the integral closedness of $R[X]$. Recall that necessary conditions for this to happen are that R be reduced and integrally closed. These conditions are not sufficient as will be seen in the example section. Suitable characterizations for $R[X]$ to be integrally closed are not known. However if $\operatorname{Min} R$ is compact, then a characterization may be given.

Theorem 13.9 Let R be a reduced integrally closed ring with $\operatorname{Min} R$ compact. Then $R[X]$ is integrally closed if and only if $T(R)$ is a von Neumann regular ring.

PROOF: ($\Leftarrow$): From Theorem 13.8, R_M is an integrally closed domain, for each $M \in \operatorname{Max} R$. Theorem 13.3 implies that $R[X]$ is integrally closed.

($\Rightarrow$): Let M be a maximal ideal in R and choose $\mathcal{M}$ in $R[X]$ such that $R \cap \mathcal{M} = M$. By Corollary 4.6, $T(R[X])$ is a von Neumann regular ring; hence $\operatorname{Min} R[X]$ is compact, Theorem 4.5. Theorem 13.8 implies that $R[X]_{\mathcal{M}}$ is an integral domain.

We show that R_M is a domain by showing that R_M may be embedded into $R[X]_{\mathcal{M}}$. As in the proof of Lemma 13.1, $R_M[X] \cong R[X]_{R\setminus M}$ and $R[X]_{\mathcal{M}}$ is a quotient ring of $R[X]_{R\setminus M}$. Let $\varphi : R[X]_{R\setminus M} \to R[X]_{\mathcal{M}}$ be the canonical homomorphism. Hence we may assume that $R_M \subseteq R[X]_{R\setminus M} \xrightarrow{\varphi} R[X]_{\mathcal{M}}$ and we must show that φ is injective on R_M. Let $a/b \in R_M$, where $b \in R \setminus M$. Suppose that $\varphi(a/b) = (a/b)/1 = 0$ in $R[X]_{\mathcal{M}}$. Then there exists some $g \in R[X] \setminus \mathcal{M}$ such that $(a/b)g = 0$ in $R[X]_{R\setminus M}$. Hence there exists some $s \in R \setminus M$ such that $sag = 0$. Since $g \notin \mathcal{M}$, some coefficient of g, say c, is in $R \setminus M$. Therefore $sca = 0$ which implies that $a/b = 0$. Complete the proof by Theorem 13.7.

The following ring construction is used in the proof of the next theorem. Let R be a ring and X an indeterminate. Define $R(X) = R[X]_S$, where $S = \{f \in R[X] : c(f) = R\}$. (Recall that $c(f)$ is the ideal of R generated by the coefficients of f.)

Theorem 13.10 Let R be a quasilocal reduced ring with maximal ideal M and let $t = g(X)/f(X)$ be an element of $T(R[X])$ that is integral over $R[X]$. If $f(X)$ has a unit coefficient, then $t \in R[X]$.

PROOF: *Case 1*: The leading coefficient of $f(X)$ is a unit. Without harm we may assume that $f(X)$ is a monic polynomial. Choose $h(X), r(X) \in R[X]$ such that $g(X) = h(X)f(X) + r(X)$, where $0 \leq \deg r(X) < \deg f(X)$. If $r(X) \neq 0$, then $r(X)/f(X)$ is integral over $R[X]$; hence it is almost integral over $R[X]$. By [G, p. 134], there is a regular element $p(X)$ in $R[X]$ such that

$p(X)(r(X)/f(X))^n$ is in $R[X]$, for each positive integer n. For each n, write $p(X)r(X)^n = f(x)^n p_n(X)$ where $p_n(X)$ belongs to $R[X]$ and depends on n. Since $f(X)$ is monic, for each n

$$\begin{aligned} n(\deg f(X)) + \deg p_n(X) & \\ = \deg f(X)^n p_n(X) & \\ = \deg p(X)r(X)^n & \\ \leq \deg p(X) + n(\deg r(X)) & \end{aligned}$$

Thus $n(\deg f(X) - \deg r(X)) \leq \deg p(X)$, for each n, a contradiction. This shows that $t = g(X)/f(X) \in R[X]$.

Case 2: The constant term of $f(X)$ is a unit. By replacing t by $t + X$, we assume that $\deg f(X) < \deg g(X)$. Let $t^n + h_1(X)t^{n-1} + \cdots + h_n(X) = 0$ be the equation of integral dependence of t over $R[X]$. Then

$$g(X)^n + h_1(X)g(X)^{n-1}f(X) + \cdots + h_n(X)f(X)^n = 0 \quad (13c)$$

Apply the R-automorphism φ of $T(R[X])$ induced by $X \mapsto 1/X$ to (13c) to get

$$g(1/X)^n + h_1(1/X)g(1/X)^{n-1}f(1/X) + \cdots + h_n(1/X)f(1/X)^n = 0 \quad (13d)$$

Let $a = \deg f(X)$, $b = \deg g(X)$, and $c = 1 + \max\{\deg h_i(X)\}$. Multiply (13d) by $X^{n(b+c)}$. Then

$$\{X^c\tilde{g}(X)\}^n + \tilde{h}_1(X)\{X^c\tilde{g}(X)\}^{n-1}\tilde{f}(X) + \cdots + \tilde{h}_n(X)\tilde{f}(X)^n = 0 \quad (13e)$$

where $\tilde{f}(X) = X^a f(1/X)$, $\tilde{g}(X) = X^b g(1/X)$, and $\tilde{h}_i(X) \in R[X]$ for $i = 1, \ldots, n$. It follows that $X^c\tilde{g}(X)/\tilde{f}(X)$ is integral over

$R[X]$. Since $\tilde{f}(X)$ has unit leading coefficient, $\{X^c\tilde{g}(X)/\tilde{f}(X)\}$ is in $R[X]$ (Case 1), so there exists $k(X) \in R[X]$ such that $X^c\tilde{g}(X) = \tilde{f}(X)k(X)$. Again, apply the R-automorphism to this last equation; then $g(X) = (X^{c+b-a})f(X)k(1/X)$. To complete the proof of this case, it is sufficient to show that $(X^{c+b-a})k(1/X)$ belongs to $R[X]$; or equivalently, to show that $\deg k(X) \leq b+c-a$. But, $\deg\{\tilde{f}(X)k(X)\} = a + \deg k(X) = c + \deg \tilde{g}(X) \leq c + b$, and so $\deg k(X) \leq c + b - a$.

Case 3. The general case. If R/M is an infinite field, then there exists an $a \in R$ such that $f(a)$ is a unit in R; for $f(X) \neq 0$ (mod M). Note that $f(X + a) = X\tilde{h}(X) + f(a)$, where $\tilde{h}(X) \in R[X]$. By Case 2, $g(X + a) = k(X)f(X + a)$ with $k(X) \in R[X]$. Consequently $g(X) = k(X - a)f(X)$, where $k(X - a) \in R[X]$.

Assume that R/M is a finite field. Let Y be an indeterminate and consider the ring $R(Y)$. From [N, Section 6], $R(Y)$ is a quasilocal ring with maximal ideal $MR(Y)$ and $R(Y)/MR(Y) \cong (R/M)(Y)$ is an infinite field. Certainly t is contained in $T(R(Y)[X])$ and is integral over $R(Y)[X]$. Thus $t \in R(Y)[X] \cap T(R[X]) = R[X]$.

Theorem 13.11 If R is an integrally closed reduced ring satisfying Property A, then $R[X]$ is integrally closed.

PROOF: It is sufficient to prove the theorem for $R = T(R)$, Corollary 13.4. Fix an $\mathcal{M} \in \operatorname{Max} R[X]$ and let $\mathcal{M} \cap R = P$. Since $R \setminus P \subseteq R[X] \setminus \mathcal{M}$, $R[X]_{\mathcal{M}}$ is a quotient ring of $R[X]_{R\setminus P}$. We have the following commutative diagram

$$\begin{array}{ccc} R[X] & \xrightarrow{\varphi'_{\mathcal{M}}} & R[X]_{\mathcal{M}} \\ {\scriptstyle\alpha'}\downarrow & & \downarrow{\scriptstyle i} \\ R[X]_{R\setminus P} & \xrightarrow[\beta']{} & R[X]_{\mathcal{M}} \end{array}$$

where i is the identity; and $\varphi'_{\mathcal{M}}$, α', and β' are canonical maps into the appropriate rings. Each map may be extended uniquely to a total quotient ring map. Then

$$\begin{array}{ccc} T(R[X]) & \xrightarrow{\varphi_{\mathcal{M}}} & T(R[X]_{\mathcal{M}}) \\ \alpha \downarrow & & \downarrow i \\ T(R[X]_{R\backslash P}) & \xrightarrow[\beta]{} & T(R[X]_{\mathcal{M}}) \end{array}$$

is commutative. Assume that $t = g/f \in T(R[X])$ and is integral over $R[X]$. By the proof of Lemma 13.2, we must show that $\varphi_{\mathcal{M}}(t) \in R[X]_{\mathcal{M}}$. By the above diagram, this amounts to showing that $\alpha(t) \in R[X]_{R\backslash P}$.

The element $\alpha(t)$ is integral over $R[X]_{R\backslash P}$. By Property A, the ideal $c(f)$ is regular in $R = T(R)$; that is, $c(f) = R$. Therefore some coefficient of f does not belong to P. But $R[X]_{R\backslash P} \cong R_P[X]$. If $\gamma : R[X]_{R\backslash P} \to R_P[X]$ is the natural isomorphism, then $\gamma\alpha(f)$ has a unit coefficient in R_P. Certainly $\gamma\alpha(g)/\gamma\alpha(f)$ is integral over $R_P[X]$, and hence in $R_P[X]$. Therefore $\alpha(t) \in R[X]_{R\backslash P}$, Theorem 13.10.

If $R[X] = R[X]'$, then R is integrally closed. Assume that $R[X] = R[X]'$ and let $a \in R'$. Then $a \in R'[X]' = R[X]$. Therefore $a \in R[X] \cap T(R) = R$.

A stronger version of Theorems 13.3 and 13.11 is proved in Section 16 (after the theory of the ring $R(X)$ is developed).

Notes

Section 9 Krull [62] proved in 1932 that the integral closure of an integral domain D is the intersection of the valuation overrings of D. Samuel [101] generalized this when he showed that the integral closure of an arbitrary commutative ring R in an extension ring T is the intersection of the dominated polynomial rings between T and R. Griffin's proof of the equivalence of paravaluation rings and dominated polynomial rings (see Notes at the end of Section 2) yields Krull's theorem for paravaluation rings. Huckaba [53] gave a short proof of this last result, when $T = T(R)$. Finally Gräter [41] gave a new and shorter proof of the same result for arbitrary T. Gräter's proof is the one given for (9.1).

Does the integral closure of R in $T(R)$ equal the intersection of the valuation overrings of R? If R is a Marot ring, the answer is yes; see (9.3). This is given in [75]. The complete answer is not known.

Section 10 That the integral closure of a Noetherian ring is a Krull ring (10.1) is due to Huckaba [54]. Ratliff [98] proved (10.2). Theorem (10.3) is proved by Portelli and Spangher [94]. They call this the strong approximation theorem for Krull rings. Nishimura [88] proved the integral domain case of (10.4). The ring theory case that is presented here is a straightforward generalization of the domain case.

Section 11 Results (11.2), (11.3), (11.4) and (11.6) appear in [54]. However there is a mistake in the proof of (11.6). (The mistake is in step 4, p. 164 of [54], where it is assumed that $R'(X)$ is the integral closure of $R(X)$ in $T(R)(X)$. See Section 16 for a discussion of this problem.) In personal correspondence D. D. Anderson gave a correct proof of (11.6). More precisely he developed the theory, including the statements and proofs of (10.4) and (11.5) that lead to the proof of (11.6). I wish to thank him for allowing me to present these results here.

Section 12 Davis [24] characterized when the integral closure of a two-dimensional Noetherian ring is Noetherian, (12.3) and (12.4). He also established much of (12.6). The versions of (12.3), (12.4), and (12.6) given in this book are from D. D. Anderson and

Huckaba [7]. Matijevic is responsible for the idea of the maximal ideal transform and for (12.5) [77].

Section 13 Akiba [1] established (13.2), (13.3), (13.9), (13.10) and (13.11). The homological results, (13.5) and (13.6), are by Cox and Pendleton [23]. Quentel [96] proved (13.7), and Endo [29] (13.8). Satisfactory necessary and sufficient conditions for $R[X]$ to be integrally closed are not known.

Chapter IV

Overrings of Polynomial Rings

The theory for overrings of polynomial rings in an arbitrary number of indeterminates is much the same as the theory for a single indeterminate. The proofs for the general case usually reduce, in easy steps, to the proofs for a single indeterminate. In this chapter we present the simpler case. There are, however, two places where we need to consider more than one indeterminate. It will be clear when these occasions arise.

14. Properties of $R(X)$

In Section 13 we defined the ring $R(X)$ to be the quotient ring $R[X]_S$ of the polynomial ring $R[X]$, where S is the set of $f \in R[X]$ such that $c(f) = R$. This ring has been quite useful in commutative ring theory. The basic properties of $R(X)$ are given in this section. When there is a possibility of ambiguity, we write $c_R(f)$ for $c(f)$.

Theorem 14.1 Let R be a ring, let $\{M_\beta\}$ be the set of maximal ideals of R, and let $S = \{f \in R[X] : c(f) = R\}$.

(1) $S = R[X] \setminus \cup M_\beta[X]$ and S is a regular multiplicatively closed subset of $R[X]$. Thus $R[X]_S = R(X)$ is an overring of $R[X]$.
(2) There is a one-to-one correspondence between the maximal (minimal) prime ideals of R and the maximal (minimal) prime ideals of $R(X)$ given by $P \leftrightarrow PR(X)$.
(3) If I is an ideal of R, then $IR(X) \cap R = I$.
(4) If Q is a P-primary ideal of R, then $QR(X)$ is $PR(X)$-primary.
(5) If I is an ideal of R, then $R(X)/IR(X)$ is isomorphic to $(R/I)(X)$.
(6) For each maximal ideal M of R, $R_M(X) = R[X]_{M[X]} = R(X)_{MR(X)}$.
(7) For each maximal ideal M of R, $R[X]_{[M[X]]} = R(X)_{[MR(X)]} = R[X]_{(M[X])} = R(X)_{(MR(X))}$.
(8) If $I_1, \ldots, I_n$ are ideals of R, then $(\cap I_i)R(X) = \cap(I_i R(X))$.

PROOF: Parts (1), (2), (3), (4), (5), and (8) can be found in either [N, p. 18] or [G, p. 410].

(6): We have $R_M(X) = R_M[X]_{S'}$, where $S' = R_M[X] \setminus MR_M[X]$. Thus $R_M[X]_{S'} = R_M[X]_{MR_M[X]} = R[X]_{M[X]} = (R[X]_S)_{M[X]_S} = R(X)_{MR(X)}$.

(7): Note that $R[X]$ and $R(X)$ are Marot rings (Theorem 7.5). By Theorem 7.6, $R(X)_{[MR(X)]} = R(X)_{(MR(X))}$ and $R[X]_{[M(X)]} = R[X]_{(M[X])}$. As in the argument of (6), $R[X]_{[M[X]]} = R(X)_{[MR(X)]}$.

Remark All of Theorem 14.1 holds for more than one indeterminate.

Theorem 14.2 For each ring R, $R(X)$ is a Marot ring with Property A, and if R is reduced, $R(X)$ has (a.c.).

PROOF: The ring $R[X]$ has these properties and each of these properties ascends to overrings.

A well-known property possessed by each commutative ring is that if $P_1, \ldots, P_n$ are prime ideals of R and if I is contained in the union of the P_i, then I is contained in a particular P_i. This has been extended to the case when only $n-2$ of the P_i's need be prime [K, p. 55]. The ring $R(X)$ satisfies a much stronger version of this result. Define R to be a *μ-ring* if I, $A_1, \ldots, A_n$ are ideals of R such that when $I \subseteq \cup A_i$, then I is contained in some A_i. We will eventually show that $R(X)$ is a μ-ring. Two lemmas are needed.

Lemma 14.3 If a ring R contains an infinite set F with the property that $x - y$ is a unit in R for all $x \neq y$ in F, then R is a μ-ring.

PROOF: Assume that I, $A_1, \ldots, A_n$ are ideals of R such that $I \subseteq \cup A_i$. If $B_i = I \cap A_i$, then $I = \cup B_i$. To show that R is a μ-ring, it suffices to show that there is some i such that $I = B_i$. We may assume that there is no i such that B_i is contained in $\cup_{j \neq i} B_j$. Suppose that $I \neq B_i$, for each i. Choose $b_i \in B_i \setminus \cup_{j \neq i} B_j$, for $i = 1, 2$. Let $E = \{b_1 + x b_2 : x \in F\}$. Then $E \subseteq I = \cup B_i$. Since F is infinite, there exist $x \neq y$ in F such that $b_1 + x b_2$ and $b_1 + y b_2$ belong to the same B_i, where $i \neq 2$. It follows that $b_2 \in B_i$, a contradiction. Therefore $I = B_i$ for an appropriate i.

Lemma 14.4 If R is a μ-ring and if N is a multiplicatively closed subset of R, then R_N is a μ-ring. If R_N is a μ-ring for each multiplicatively closed subset N of R, where N is the complement of a finite union of maximal ideals, then R is a μ-ring.

PROOF: The first statement is clear, so we concentrate on the second. Let I, $A_1, \ldots, A_n$ be ideals of R such that $I \subseteq \cup A_i$ and let $B_i = I \cap A_i$. Then $I = \cup B_i$. Assume that $I \supset B_i$ for each i. Choose $x_i \in I \setminus B_i$ and let M_i be a maximal ideal of R containing $(B_i :_R x_i)$. If $N = R \setminus \cup_{i=1}^n M_i$, then $IR_N = \cup(B_i R_N)$. But $x_i/1 \in IR_N \setminus B_i R_N$ for each i, contradicting the hypothesis.

Theorem 14.5 For each ring R, $R(X)$ is a μ-ring.

PROOF: If $M_1, \ldots, M_n$ are maximal ideals of R, then $M_1R(X)$, $\ldots, M_nR(X)$ are maximal ideals of $R(X)$. In view of Lemma 14.4, we need to prove that $R(X)_N$ is a μ-ring, where $N = R(X) \setminus \cup M_i R(X)$. The maximal ideals of $R(X)_N$ are $M_1R(X)_N, \ldots,$ $M_nR(X)_N$. For each i,

$$\begin{aligned} R(X)_N/M_iR(X)_N &\cong ((R(X)/M_iR(X))_{(N+M_iR(X))/M_iR(X)} \\ &\cong ((R/M_i)(X))_{N'} \end{aligned}$$

for an appropriate multiplicatively closed set N'. This shows that for each i the residue field $R(X)_N/M_iR(X)_N$ is infinite. Therefore it suffices to prove the theorem for a ring R with finitely many maximal ideals $M_1, \ldots, M_n$ such that each R/M_i is infinite.

Let R be such a ring. Let I, $A_1, \ldots, A_t$ be ideals of R such that $I \subseteq \cup A_i$. If $B_i = I \cap A_i$, then $I = \cup B_i$ and we must prove that $I = B_i$ for some i. Let F_i be a complete set of representatives of the nonzero elements of R/M_i; that is, for each nonzero $\bar{x} \in R/M_i$, F_i contains exactly one element s such that $s + M_i = \bar{x}$. Let $\{s_{ij} : j = 1, 2, \ldots\}$ be a denumerable subset of F_i, for $i = 1, 2, \ldots, n$. By the Chinese remainder theorem [ZSI, p. 177], for each j there exists some $x_j \in R$ such that $x_j \equiv s_{ij} \pmod{M_i}$, $i = 1, \ldots, n$. Let F be the set of all x_j obtained in this fashion. Then F is an infinite set satisfying the hypotheses of Lemma 14.3. Therefore R is a μ-ring.

The idempotent elements of R and $R(X)$ coincide. We need Lemma 14.6 as a prelude.

Lemma 14.6 If f is an element of the polynomial ring $R[X]$ with $c(f) = (a)$, then there is a $g \in R[X]$ such that $f = ag$ and $c(g) = R$.

PROOF: Let $f = a_0 + a_1X + \cdots + a_nX^n$, let $a_i = r_ia$, and let $a = \sum_{i=0}^{n} s_ia_i$ where $r_i, s_i \in R$. If $d = \sum_{i=0}^{n} s_ir_i$, then $a = da$, and thus $(1-d)a = 0$. The required element is $g = r_0 + r_1X + \cdots + r_nX^n + (1-d)X^{n+1}$.

Theorem 14.7 If R is a ring, then the idempotent elements of R and $R(X)$ coincide.

PROOF: We must show that if an idempotent element is in $R(X)$, then it is in R. Let f/g be an idempotent of $R(X)$ where $f, g \in R[X]$ and $c(f) = R$. Then $fg^2 = gf^2$ implies that $fg = f^2$, since g is a regular element in $R[X]$. By the content formula [G, p. 343], $c(g)^{k+1}c(f) = c(g)^kc(fg)$ for some nonnegative integer k. Since $c(g) = R$, $c(f) = c(fg)$. Hence $c(f) = c(fg) = c(f^2) \subseteq c(f)^2$, and thus $c(f) = c(f)^2$. But a finitely generated idempotent ideal is principal and generated by an idempotent element [G, p. 63]. So $c(f) = (e)$ where e is idempotent in R. By Lemma 14.6, there is a polynomial f' in $R[X]$ such that $f = ef'$ and $c(f') = R$. Then $ef'/g = f/g = f^2/g^2 = e^2(f')^2/g^2$. Since f' and g are regular, and since e is idempotent, $e = ef'/g = f/g \in R$.

15. Principal Ideals and Pic $R(X)$

Two closely related ideas for a ring R are the behavior of locally principal ideals of R and the Picard group of R denoted by Pic R. An ideal I of R is said to be *locally principal* if $IR_M = I_M$ is a principal ideal for each maximal ideal M of R. We show that if I is a locally principal ideal of $R(X)$, then I is principal. For the integral domain case this would be enough to prove that $\operatorname{Pic} R(X) = 0$. This is, in fact, as we shall prove, true for an arbitrary ring R; that is, for each ring R, $\operatorname{Pic} R(X) = 0$. We begin with some results on locally principal ideals.

Theorem 15.1 Let R be a ring and let $f \in R[X]$. The following statements are equivalent:

(1) $c(f)$ is locally principal.
(2) $fR(X) = c(f)R(X)$.
(3) $fR(X) = IR(X)$, for some ideal I of R.
(4) $c(f)R(X)$ is principal.
(5) $c(f)R(X)$ is locally principal.

PROOF: (1) $\Rightarrow$ (2): We need to prove that $fR(X)_{MR(X)} = c(f)R(X)_{MR(X)}$, for each maximal ideal $MR(X)$ in $R(X)$. By Theorem 14.1(6), this is equivalent to proving that $fR_M(X) = c(f)R_M(X)$, for each $M \in \operatorname{Max} R$. Therefore we may assume that R is a quasilocal ring. If $f = a_0 + a_1X + \cdots + a_nX^n$ is in $R[X]$, then there is some $b \in R$ such that $c(f) = (b)$. Write $b = \sum r_ia_i$ and $a_i = s_ib$ where $r_i, s_i \in R$. The equation $(1 - \sum r_is_i)b = 0$ implies that some s_t is a unit of R. Thus $(a_t) = (b) = c(f)$. Write $a_i = d_ia_t$ and let $h = d_0 + d_1X + \cdots + d_nX^n$, where $d_i \in R$ and $d_t = 1$. Then $c(h) = R$; whence h is a unit in $R(X)$. Therefore $c(f)R(X) = a_tR(X) = a_thR(X) = fR(X)$.

(2) $\Rightarrow$ (3): Clear.

(3) $\Rightarrow$ (1): Note that $IR(X) = fR(X) \subseteq c(f)R(X)$; thus $I \subseteq c(f)$. Without loss of generality assume that R is a quasilocal ring. We must show that $c(f)$ is a principal ideal. Choose $b_i \in I, f_i \in R[X]$ and $g \in R[X] \setminus M[X]$ such that $f = \sum b_if_i/g$. For each i, write $b_i = fh_i/h$ where $h_i \in R[X]$ and $h \in R[X] \setminus M[X]$. Thus $1 - \sum h_if_i/gh \in MR(X)$, since it is a nonunit in the quasilocal ring $R(X)$. It follows that some h_j is a unit in $R(X)$. Therefore $f = b_jh/h_j$, which implies that $fR(X) = b_jR(X)$. By [G, p. 343]

$c(f) = c(f)c(h_j) = c(fh_j) = c(b_jh) = b_jc(h) \subseteq (b_j) \subseteq c(f)$. Therefore $c(f) = (b_j)$ is a principal ideal.

(2) $\Rightarrow$ (4) $\Rightarrow$ (5): Clear.

(5) $\Rightarrow$ (1): Again we may assume that R is a quasilocal ring. Thus $c(f)R(X) = aR(X)$ for some $a \in c(f)$. Write $f = ag/h$ where $g, h \in R[X]$ and $c(h) = R$. As in (3) $\Rightarrow$ (1), $c(f) = (a)$.

Corollary 15.2 Let I be an ideal of the ring R, then the following are equivalent:

(1) I is finitely generated and locally principal.
(2) $IR(X)$ is finitely generated and locally principal.
(3) $IR(X)$ is principal.

PROOF: (1) $\Rightarrow$ (3): Suppose that $I = (a_0, \ldots, a_n)$. Let $f = a_0 + a_1X + \cdots + a_nX^n$; then $I = c(f)$. By Theorem 15.1, $IR(X)$ is principal.

(3) $\Rightarrow$ (2): Obvious.

(2) $\Rightarrow$ (1): Since $IR(X)$ is finitely generated, there is a finitely generated ideal I' of R such that $IR(X) = I'R(X)$. Contracting back to R yields $I = I'$. Then $I = c(f)$ for some $f \in R[X]$. By Theorem 15.1, $I = I' = c(f)$ is locally principal.

Let X and Y be two indeterminates over a ring R. Define $R(X,Y)$ as $R[X,Y]_{S_1}$ where $S_1 = \{f \in R[X,Y] : c_R(f) = R\}$. We know that $R(X) = R[X]_S$ where $S = \{f \in R[X] : c_R(f) = R\}$. Define $R(X)(Y) = R(X)[Y]_{S_2}$ where $S_2 = \{k \in R(X)[Y] : c_{R(X)}(k) = R(X)\}$.

Lemma 15.3 With the same notation as above, $R(X)(Y) = R(X,Y) = R(Y,X) = R(Y)(X)$.

PROOF: By symmetry it suffices to prove that $R(X)(Y) = R(X,Y)$. Let $\{M_\beta\}$ be the set of maximal ideals of R and let $f/g \in R(X,Y)$ such that $f \in R[X,Y]$ and $g \in S_1 = R[X,Y] \setminus \cup M_\beta R[X,Y]$. For each maximal ideal M of R, $MR(X)[Y] =$

$MR[X]_S[Y] = MR[X][Y]_S = MR[X,Y]_S$. Since $MR[X,Y]$ is a prime ideal of $R[X,Y]$ disjoint from S, $MR(X)[Y] \cap R[X,Y] = MR[X,Y]_S \cap R[X,Y] = MR[X,Y]$. Hence $g \in R(X)[Y] \setminus \cup M_\beta R(X)[Y]$, and thus $f/g \in R(X)(Y)$.

For the other inclusion let $f/g \in R(X)(Y)$, where $f \in R(X)[Y]$ and $g \in S_2 = R(X)[Y] \setminus \cup M_\beta R(X)[Y]$. Write $f = f'/h$ and $g = g'/k$ where $f' \in R[X,Y]$, $g' \in S_1 = R[X,Y] \setminus \cup M_\beta [X,Y]$, and h, k are in $S = R[X] \setminus \cup M_\beta[X]$. Then $f/g = f'k/hg' \in R(X,Y)$.

Theorem 15.4 Each finitely generated locally principal ideal of the ring $R(X)$ is principal.

PROOF: Let $I = (f_0, \ldots, f_n)R(X), f_i \in R[X]$, be a locally principal ideal of $R(X)$. If $t = 1 + \max\{\deg f_i\}$ and if $f = f_0 + X^t f_1 + X^{2t} f_2 + \cdots + X^{nt} f_n$, then $c(f) = c(f_0) + c(f_1) + \cdots + c(f_n)$. Passing to the ring $R(X,Y) = R(X)(Y)$, we see that $IR(X,Y)$ is a principal ideal (Corollary 15.2). If $g = f_0 + f_1 Y + \cdots + f_n Y^n$, then $c_{R(X)}(g) = I$. By Theorem 15.1, $IR(X,Y) = gR(X,Y)$, so $f = gh/k$ where $h, k \in R[X,Y]$ and $c(k) = R$. Using the content formula [G, p. 343], we have

$$c(f) = c(k)c(f) = c(kf) = c(gh) \subseteq c(g)c(h) = c(f)c(h)$$

Therefore $c(f) = c(f)c(h)$.

Let M be a maximal ideal of R; then $c(f)_M = c(f)_M c(h)_M$. By Nakayama's lemma [K, p. 51], either $c(f)_M = (0)_M$ or $c(h)_M = R_M$. If $c(f)_M = (0)_M$, then $IR_M(X) = (0) = fR_M(X)$. If $c(h)_M = R_M$, then the image of h is a unit in the ring $R_M(X,Y)$. Hence $fR_M(X,Y) = gR_M(X,Y)$. Therefore

$$\begin{aligned} fR_M(X) &= fR_M(X)(Y) \cap R_M(X) \\ &= gR_M(X)(Y) \cap R_M(X) \\ &= IR_M(X)(Y) \cap R_M(X) \end{aligned}$$

$$= IR_M(X)$$

Using the fact that $R_M(X) = R(X)_{MR(X)}$ we see that f and I are locally equal, hence equal.

We are ready to discuss the Picard group of $R(X)$. We first summarize some results given in Section 5, Chapter II of [B]. Let $\mathcal{P}$ be a finitely generated projective module over the ring R. If P is a prime ideal of R, then $\mathcal{P}_P$ is a free R_P-module. Let n be a nonnegative integer. A projective R-module $\mathcal{P}$ has *rank* n if it is finitely generated and the rank of the R_P-module $\mathcal{P}_P$ is n, for each $P \in \operatorname{Spec} R$. If $\mathcal{P}$ is a finitely generated projective rank one R-module, let $[\mathcal{P}]$ denote the isomorphism class of $\mathcal{P}$ (as R-modules) and let $\operatorname{Pic} R$ be the set of all such classes. If $[\mathcal{P}]$, $[\mathcal{Q}] \in \operatorname{Pic} R$, the $\mathcal{P} \otimes_R \mathcal{Q}$ is again a finitely generated projective rank one R-module. Thus $[\mathcal{P} \otimes_R \mathcal{Q}]$ is in $\operatorname{Pic} R$. Under the operation $[\mathcal{P}] + [\mathcal{Q}] = [\mathcal{P} \otimes_R \mathcal{Q}]$, $\operatorname{Pic} R$ is an abelian group, called the *Picard group* of R. The identity of $\operatorname{Pic} R$ is the equivalence class of free rank one R-modules. Let R be an integral domain, let $\mathcal{I}$ be the group of invertible fractional ideals of R, and let $\mathcal{K}$ be the group of principal fractional ideals of R. Then the *class group* of R, denoted by $\operatorname{Cl} R$ is defined by $\operatorname{Cl} R = \mathcal{I}/\mathcal{K}$. It is well known that in the integral domain case that $\operatorname{Pic} R$ and $\operatorname{Cl} R$ are isomorphic [B, p. 120].

If D is an integral domain, then so is $D(X)$. If I is an invertible fractional ideal in $D(X)$, then I is locally principal [K, p. 38]. Thus I is principal (Theorem 15.4). This shows that $\operatorname{Pic} D(X) = (0)$. The remainder of this section is devoted to proving that $\operatorname{Pic} R(X) = (0)$, for any commutative ring R.

We sketch some elementary results concerning R-modules. Let $\mathcal{A}$ be an R-module. Define $\mathcal{A}[X]$ to be the set of all elements of the form $a_0 + a_1 X + \cdots + a_n X^n$ where $a_i \in \mathcal{A}$. Define the product of an element of $R[X]$ and an element of $\mathcal{A}[X]$ in the obvious way. Then $\mathcal{A}[X]$ becomes an $R[X]$-module. For a multiplicatively closed set S of $R[X], \mathcal{A}[X]_S$ is an $R[X]_S$-module. When $S = \{f \in R[X] : c(f) = R\}$, we denote the $R[X]_S = R(X)$-module $\mathcal{A}[X]_S$ by $\mathcal{A}(X)$.

Assume that $\mathcal{P}$ is a finitely generated projective rank one R-module. A routine argument shows that $\mathcal{P}(X)$ is a finitely generated projective $R(X)$-module. We claim that $\mathcal{P}(X)$ is also rank one. It is sufficient to show that if $\mathcal{M}$ is a maximal ideal of $R(X)$, then $\mathcal{P}(X)_{\mathcal{M}}$ is a free rank one $R(X)_{\mathcal{M}}$-module. Let $\mathcal{M} \cap R = M$. The mapping $\mathcal{P}_{\mathcal{M}}[X] \to \mathcal{P}[X]_{R \setminus M}$ given by $p_0/b + (p_1/b)X + \cdots + (p_n/b)X^n \mapsto (p_0 + p_1X + \cdots + p_nX^n)/b$ is an isomorphism. Using this fact and modifying the proof of Theorem 14.1(6), we see that $\mathcal{P}_M(X) = \mathcal{P}(X)_{\mathcal{M}}$. Since $\mathcal{P}_M$ is a rank one free R_M-module, $\mathcal{P}_M(X)$ is a rank one free $R_M(X)$-module; that is, $\mathcal{P}(X)_{\mathcal{M}}$ is a rank one free $R(X)_{\mathcal{M}}$-module.

Let $\mathcal{A}$ be an R-module and $\mathcal{N}$ an R-submodule of $\mathcal{A}$, then $R(X)\mathcal{N} = \mathcal{N}(X)$ and $\mathcal{N}(X) \cap \mathcal{A} = \mathcal{N}$. A locally cyclic R-module is an R-module $\mathcal{A}$ such that $\mathcal{A}_{\mathcal{M}}$ is a cyclic R_M-module for each maximal ideal M.

Lemma 15.5 If $\mathcal{A}$ is a finitely generated locally cyclic R-module, then $\mathcal{A}(X)$ is a cyclic $R(X)$-module. Moreover if $\{a_0, \ldots, a_n\}$ spans $\mathcal{A}$, then $f = a_0 + a_1X + \cdots + a_nX^n$ generates $\mathcal{A}(X)$.

PROOF: Clearly $R(X)f \subseteq \mathcal{A}(X)$. We show that $\mathcal{A}(X)$ and $R(X)f$ are locally equal. Let M be a maximal ideal of R; then $\mathcal{A}_M$ is a cyclic R_M-module. Nakayama's lemma says that $MR_M\mathcal{A}_M \subset \mathcal{A}_M$ or $\mathcal{A}_M = (0)_M$. If the latter holds, then $(0)_M = \mathcal{A}_M(X) = R_M(X)(f/1)$; so we concentrate on the former. Hence there exists some j such that $a_j/1 \in \mathcal{A}_M \setminus MR_M\mathcal{A}_M$ ($a_j/1$ is the image of a_j in $\mathcal{A}_M$). Then $\mathcal{A}_M/MR_M\mathcal{A}_M$ is a one-dimensional vector space over R_M/MR_M and $a_j/1 + MR_M\mathcal{A}_M$ may be taken as a basis. It follows that $\mathcal{A}_M = R_M(a_j/1) + MR_M\mathcal{A}_M$. By [N, p. 12], $\mathcal{A}_M = R_M(a_j/1)$. Thus $a_i/1 = r_ia_j/1$ where $r_i \in R_M$ and $r_j = 1/1$. Let $h = r_0 + r_1X + \cdots + r_nX^n$, so $c_{R_M}(h) = R_M$. Since $a_0/1 + (a_1/1)X + \cdots + (a_n/1)X^n = ha_j/1$, $\mathcal{A}_M(X) =$

$R_M(X)(f/1)$. Since $\mathcal{A}(X)$ and $R(X)f$ are locally equal, they are equal.

Lemma 15.6 If $\mathcal{P}$ is a projective rank one cyclic R-module, then $\mathcal{P}$ is free.

PROOF: We have a split exact sequence $0 \to K \to R \to \mathcal{P} \to 0$. Hence $R = \mathcal{P} \oplus K$ where we can assume that $\mathcal{P}$ and K are ideals of R. Assume that $K \neq (0)$. If P is a minimal prime ideal of $\mathcal{P}$; that is, minimal with respect to containing $\mathcal{P}$, then $\mathcal{P}_P = (0)$, contradicting the hypothesis that $\mathcal{P}$ is projective rank one R-module. Therefore $K = (0)$ and hence $\mathcal{P}$ is free.

We are ready to show that $\operatorname{Pic} R(X) = (0)$. The proof given here is based on the proof of Theorem 15.4.

Theorem 15.7 For each commutative ring R, $\operatorname{Pic} R(X) = (0)$.

PROOF: Let $\mathcal{P}$ be a finitely generated rank one projective $R(X)$-module. We need to show that $\mathcal{P}$ is free. We can assume that $\mathcal{P}$ is contained in the n-dimensional free $R(X)$-module $R(X)^n$. Choose a set of generators $P_0, \ldots, P_m$ of $\mathcal{P}$. For $i = 0, \ldots, m$ write P_i as the n-tuple $(f_{i1}, \ldots, f_{in})$ where $f_{ij} \in R[X]$. Let $t = 1 + \max\{\deg f_{ij}\}$ and set $P = P_0 + X^t P_1 + \cdots + X^{tm} P_m$. Then $P \in R[X]^n$. Adopt the following notation: If G is the n-tuple $(g_1, \ldots, g_n)$ in $R[X]^n$, then $\tilde{c}(G) = (c(g_1), \ldots, c(g_n)) \subseteq R^n$. The $c(g_i)$'s are, as usual, the content ideal of the g_i's. With this notation $\tilde{c}(P) = (c(f_1), \ldots, c(f_n))$, where $f_j = f_{0j} + X^t f_{1j} + \cdots + X^{tm} f_{mj}$. By the choice of $t, c(f_j) = c(f_{0j}) + \cdots + c(f_{mj})$.

Let Y be another indeterminate and consider the ring $R(X)(Y) = R(X,Y)$, Lemma 15.3. By Lemmas 15.5 and 15.6, $\mathcal{P}(Y)$ is a rank one free $R(X,Y)$-module generated by $Q = P_0 + P_1 Y + \cdots + P_m Y^m$. Hence $P \in \mathcal{P}(Y) = R(X,Y)Q$. Let $P = (h/k)Q$ for some $h, k \in R[X,Y]$ with $c(k) = R$. By the content formula [G, p. 343] and the fact that each component of $\tilde{c}(Q)$

and $\tilde{c}(P)$ are ideals in R, we have

$$\tilde{c}(P) = c(k)\tilde{c}(P) = \tilde{c}(kP) = \tilde{c}(hQ) \subseteq c(h)\tilde{c}(Q)$$

Clearly $\tilde{c}(P) = \tilde{c}(Q)$, thus $\tilde{c}(P) = c(h)\tilde{c}(P)$, and therefore $c(f_j) = c(h)c(f_j)$ for $j = 1, \ldots, n$. Let M be a fixed maximal ideal of R. Then $c(f_j)_M = c(h)_M c(f_j)_M$ for each j. By Nakayama's lemma, either $c(f_j)_M = (0)$ for all j, or $c(h)_M = R_M$.

Assume that $c(f_j)_M = (0)$ for each j. Then $\tilde{c}(P)_M = \tilde{c}(Q)_M = ((0), \ldots, (0))$ where each (0) is the zero ideal in R_M. But $\mathcal{P}(Y) = R(X,Y)Q$ where

$$Q = (f_{01} + f_{11}Y + \cdots + f_{m1}Y^m, \ldots, \\ f_{0n} + f_{1n}Y + \cdots + f_{mn}Y^m)$$

Thus $\mathcal{P}(Y)_{MR(X,Y)}$ is the $R(X,Y)_{MR(X,Y)} = R_M(X,Y)$-module generated by the image of Q in $\mathcal{P}(X)_{MR(X,Y)}$; more precisely, $\mathcal{P}(Y)_{MR(X,Y)}$ is generated by

$$(\bar{f}_{01} + \bar{f}_{11}Y + \cdots + \bar{f}_{m1}Y^m, \ldots, \\ \bar{f}_{0n} + \bar{f}_{1n}Y + \cdots + \bar{f}_{mn}Y^m)$$

where $\bar{f}_{ij}$ is the polynomial derived from f_{ij} by reducing the coefficients modulo M. Therefore, since each $c(f_{ij})_M = (0)$, $\mathcal{P}(Y)_{MR(X,Y)} = (0)$. This cannot happen because $\mathcal{P}(Y)$ is a free rank one projective module. Hence $c(h)_M = R_M$ for each maximal ideal M of R, so $c(h) = R$, and thus h is a unit in $R(X,Y)$. Therefore $R(X,Y)P = R(X,Y)Q = \mathcal{P}(Y)$. Finally $R(X)P = (R(X,Y)(R(X)P)) \cap \mathcal{P} = \mathcal{P}(Y) \cap \mathcal{P} = \mathcal{P}$. By Lemma 15.6 $\mathcal{P}$ is free.

The content formula involving finitely generated locally principal ideals is especially nice; compare with the corresponding formula for invertible ideals.

Theorem 15.8 If I is a finitely generated locally principal ideal of a ring R and if $I = c(f)$ for some $f \in R[X]$, then $c(fg) = c(f)c(g)$ for each $g \in R[X]$.

PROOF: Let $f = a_0 + a_1X + \cdots + a_nX^n$ where $I = (a_0, \ldots, a_n)$. If M is a maximal ideal of R, then $c(f)_M = (b)_M$ for some $b \in I$. Writing $x/1$ for the image of x in R_M, we have $b/1 = \sum r_ia_i/1$ and $a_i/1 = s_ib/1$ where $r_i, s_i \in R_M$. Thus $(b/1)(1/1 - \sum r_is_i) = 0/1$, so some s_t is a unit in R_M. Consequently $(a_t)_M = (b)_M = c(f)_M$. From Lemma 14.6, we have $f/1 = (a_t/1)f_1$ where $f_1 \in R_M[X]$ and $c_{R_M}(f_1) = R_M$.

Let $g \in R[X]$; then $c(fg)_M = c_{R_M}((f/1)(g/1)) = (a_t/1)\ c_{R_M}\ ((f_1)(g/1)) = (a_t/1)\ c_{R_M}\ (f_1)\ c_{R_M}\ (g/1) = c_{R_M}(f/1)c_{R_M}(g/1) = c(f)_Mc(g)_M$. Since this is true for all M, $c(fg) = c(f)c(g)$.

16. When $R(X)$ Is Integrally Closed

If $R[X]$ is an integrally closed ring, so is $R(X)$, since $R(X)$ is a regular quotient ring of $R[X]$. There are many instances when $R(X)$ is integrally closed, but $R[X]$ is not. This section studies when $R(X)$ is integrally closed.

We recall some facts about multiplicatively closed subsets of a ring R; see Chapter 1 of [G] for the details. The *saturation* of a multiplicatively closed subset N of R is $N^* = \{x \in R : x$ is a unit in $R_N\}$. The set N is saturated if and only if $N = N^*$, and this occurs if and only if $N = R \setminus \cup P_i$ where P_i is the set of prime ideals of R disjoint from N. If N^* is the saturation of N, then N^* is the largest multiplicatively closed subset of R containing N for which $R_N = R_{N^*}$.

Theorem 16.1 Let T be an extension ring of the ring R and let R' be the integral closure of R in T. Then:

(1) $R'(X) = R'[X]_S$ where $S = \{f \in R[X] : c(f) = R\}$.

(2) $R'(X)$ is the integral closure of $R(X)$ in $T[X]_S$.

PROOF: Let $\{M_\beta\}$ and $\{M'_\gamma\}$ be the sets of maximal ideals of R and R', respectively. Let S be as above. Then $S = R[X] \setminus \cup M_\beta[X]$. Define $S' = R'[X] \setminus \cup M'_\gamma R'[X]$; then $R(X) = R[X]_S$ and $R'(X) = R'[X]_{S'}$. To prove (1) it suffices to show that S' is the saturation of S in $R'[X]$. Denote this saturation by S^*. Since $S \subseteq S'$ and S' is a saturated set, $S^* \subseteq S'$. Write $S^* = R'[X] \setminus \cup \mathcal{P}_i$ [G, p. 16], where $\{\mathcal{P}_i\}$ is a set of prime ideals of $R'[X]$. If $\mathcal{P}$ is one of the $\mathcal{P}_i$, let $\mathcal{P} \cap R[X] = \mathcal{P}_0$. Since $\mathcal{P}_0 \cap S = \emptyset, \mathcal{P}_0 \subseteq \cup M_\beta[X]$, and hence $\mathcal{P}_0 \subseteq M_\beta[X]$ for some β. We know that $R'[X]$ is integral over $R[X]$ [G, p. 96]. By the Going Up theorem [K, p. 29], there exists a prime ideal $\mathcal{Q}$ of $R'[X]$ such that $\mathcal{Q} \supseteq \mathcal{P}$ and $\mathcal{Q} \cap R[X] = M_\beta[X]$. Since $\mathcal{Q} \cap R = M_\beta$ and since R' is integral over R, there is some γ such that $\mathcal{Q} \cap R' = M'_\gamma$. Clearly $\mathcal{Q} \supseteq M'_\gamma[X]$. Then $\mathcal{Q} \cap R[X] = M'_\gamma[X] \cap R[X] = M_\beta[X]$. Since $R'[X]$ is integral over $R[X], \mathcal{Q} = M'_\gamma[X]$. Thus $\mathcal{P} \subseteq M'_\gamma[X]$. But $M'_\gamma[X] \cap S^* = \emptyset$ implies that $\mathcal{P} = M'_\gamma[X]$. Therefore $\{\mathcal{P}_i\} \subseteq \{M'_\gamma[X]\}$. This proves that $S^* = S'$.

(2): We know that $R'[X]$ is the integral closure of $R[X]$ in $T[X]$ [G, p. 96] and that $R'[X]_S$ is the integral closure of $R[X]_S$ in $T[X]_S$. Now apply (1).

Lemma 16.2 Let R be a ring and let t be an element of $R(X)$ that is integral over $R[X]$. Then there exists an element $h \in R[X]$ such that $t - h$ is nilpotent in $R(X)$.

PROOF: Let N be the nilradical of R and let $t = g/f$ where $f, g \in R[X]$ and $c(f) = R$. Define $\bar{t} = \bar{g}/\bar{f}$ where $\bar{f}$ and $\bar{g}$ are the residues of f and g in $(R/N)[X] \cong R[X]/N[X]$.

We claim that $\bar{t}$ is in $(R/N)[X]$. By the proof of Lemma 13.2, it suffices to show that $\varphi_{\mathcal{M}}(\bar{t}) \in (R/N)[X]_{\mathcal{M}}$ for each maximal ideal $\mathcal{M}$ of $(R/N)[X]$, ($\varphi_{\mathcal{M}}$ is the map defined at the beginning of Section 13.) Note that $(R/N)[X]_{\mathcal{M}}$ is a quotient ring of $(R/N)[X]_{(R/N)\backslash M} \cong (R/N)_M[X]$, where $M = \mathcal{M} \cap (R/N)$. Clearly $\bar{t}$ is integral over $(R/N)[X]$ and $c_{R/N}(\bar{f}) = R/N$. Hence some coefficient of $\bar{f}$ is not in M/N. Define $\sigma_0 : R/N \to (R/N)_M$ to be the natural homomorphism, extend this to $\sigma_1 : (R/N)[X] \to (R/N)_M[X]$ in the usual way, and let σ be the homomorphism of $T((R/N)[X]) \to T((R/N)_M[X])$ which is the unique extension of σ_1. Some coefficient of $\sigma(\bar{f})$ is a unit in $(R/N)_M$, and by [G, p. 91] $\sigma(\bar{t})$ is integral over $(R/N)_M[X]$. By Theorem 13.10 $\sigma(\bar{t}) \in (R/N)_M[X]$. If we let γ be the extension of the isomorphism of $(R/N)_M[X] \to (R/N)[X]_{(R/N)\backslash M}$ to $T((R/N)_M[X]) \to T((R/N)[X])_{R/N)\backslash M})$ and let ρ be the extension of the natural mapping $(R/N)[X]_{(R/N)\backslash M} \to (R/N)[X]_{\mathcal{M}}$ (remember that $(R/N)[X]_{\mathcal{M}}$ is a quotient ring of $(R/N)[X]_{(R/N)\backslash M}$ to $T((R/N)[X]_{(R/N)\backslash M}) \to T((R/N)[X]_{\mathcal{M}})$. This yields the following commutative diagram.

$$\begin{array}{ccc} T((R/N)[X]) & \xrightarrow{\varphi_{\mathcal{M}}} & T((R/N)[X]_{\mathcal{M}}) \\ \sigma \downarrow & & \uparrow \rho \\ T((R/N)_M[X]) & \xrightarrow[\gamma]{} & T((R/N)[X]_{(R/N)\backslash \mathcal{M}}) \end{array}$$

Therefore $\sigma(\bar{t}) \in (R/N)_M[X]$ implies that $\varphi_{\mathcal{M}}(\bar{t}) \in (R/N)[X]_{\mathcal{M}}$. This proves the claim.

Since $\bar{t}$ is in $R[X]/N[X]$, $\bar{t} = h + N[X]$ for some $h \in R[X]$. Thus $\bar{g} = \bar{f}\bar{t}$ implies that $g - fh = n$, where $n \in N[X]$. Therefore $g/f - h = n/f$ is a nilpotent element of $R(X)$.

We have used, on several occasions, the fact that if $R \subset T$ are rings and if R' is the integral closure of R in T, then $R'[X]$ is the integral closure of $R[X]$ in $T[X]$. This is not true for rings of the type $R(X)$; see Section 27. The best general statement that can be made in this regard is Theorem 16.1(2). However for reduced rings we have an analog to the polynomial ring situation.

Theorem 16.3 Let T be a reduced extension ring of the ring R and let R' be the integral closure of R in T. Then $R'(X)$ is the integral closure of $R(X)$ in $T(X)$.

PROOF: By Theorem 16.1, $R'(X)$ is integral over $R(X)$. Hence we show that if t is an element in $T(X)$ that is integral over $R(X)$, then $t \in R'(X)$. Write $t^n + (h_1/f)t^{n-1} + \cdots + h_n/f = 0$ where $h_i, f \in R[X]$ and $c(f) = R$. Then ft is integral over $R[X]$ and hence over $T[X]$. By Lemma 16.2, $ft \in T[X]$. Thus $ft \in R'[X]$ which implies that $t \in R'[X]_S = R'(X)$ (Theorem 16.1).

We shall presently see that there are other situations where the conclusion of Theorem 16.3 holds. It is now time to investigate the problem stated in the title of this section: When $R(X)$ is integrally closed. An important tool in this analysis is the following characterization of Property A.

Theorem 16.4 A ring R has Property A if and only if $T(R)(X) = T(R[X])$.

PROOF: ($\Rightarrow$): Since R has Property A, so does $T(R)$. If g is a regular element in $T(R)[X]$, then $c(g)$ is a regular ideal of $T(R)$; that is, $c(g) = T(R)$. Therefore $T(R[X]) = T(R)(X)$.

($\Leftarrow$): Let $I = (a_0, a_1, \ldots, a_n)$ be an ideal of R and define $g = a_0 + a_1X + \cdots + a_nX^n \in R[X]$. If $\operatorname{Ann} I = (0)$, then g is regular in $R[X]$ and hence $1/g \in T(R[X]) = T(R)(X)$. There exist $f, h \in T(R)[X]$ with $c_{T(R)}(h) = T(R)$ such that $f/h = 1/g$. Hence $T(R) = c_{T(R)}(h) = c_{T(R)}(fg) \subseteq c_{T(R)}(f)c_{T(R)}(g)$. Thus $c_{T(R)}(g) = T(R)$, which implies that $c_R(g)$ is a regular ideal of R. Consequently R has Property A.

The following argument parallels that given in (1) $\Rightarrow$ (2) of the proof of Theorem 12.2.

Lemma 16.5 Let R be an integrally closed ring in which the nilradical is finitely generated. Let M be a regular maximal ideal of R. Then R_M is a reduced ring.

PROOF: Note that R_M is integrally closed in $T(R)_{R\setminus M}$ and that $N(R_M) = N(R)R_M$ is finitely generated. Let b be a regular element belonging to M and let $b/1$ be the image of b under the natural homomorphism $R \to R_M$. Then $b/1$ is contained in the Jacobson radical of R_M, and is also a unit in $T(R)_{R\setminus M}$. Since $(b/1)N(T(R)_{R\setminus M}) = N(T(R)_{R\setminus M})$ and since $N(T(R)_{R\setminus M}) = N(R_M)$, we have $(b/1)N(R_M) = N(R_M)$. By Nakayama's lemma, $N(R_M) = (0)$.

Theorem 16.6 Let R be an integrally closed ring with Property A. If the nilradical of R is finitely generated, then $R(X)$ is an integrally closed ring.

PROOF: In view of Theorem 16.4 it is enough to show that if $t \in T(R)(X)$ and is integral over $R(X)$, then t is in $R(X)$. For such a t, there exists $f \in R[X]$ such that $c(f) = R$ and tf is integral over $R[X]$. Since tf is in $T(R)(X)$ and is also integral over $T(R)[X]$, there is some $h \in T(R)[X]$ such that $tf - h$ is a nilpotent element of $T(R)(X) = T(R[X])$ (Lemma 16.2). Write $tf - h = (n_0 + n_1X + \cdots + n_mX^m)/g$ where $g = a_0 + a_1X + \cdots + a_rX^r \in R[X]$ and $n_i \in N(R)$. If M is a regular maximal ideal of R, then $\text{Ann}(n_i) \not\subseteq M$, Lemma 16.5. On the other hand Property A implies that $c(g)$ is a regular ideal of R. Therefore $c(g) + \text{Ann}(n_i) = R$. Choose $b \in \text{Ann}(n_i)$ such that $(a_0, \ldots, a_r, b) = R$. Setting $g_1 = b + a_0X + \cdots + a_rX^{r+1}$, we get $n_iX^i/g = n_iX^{i+1}/g_1$, where $c(g_1) = R$. Thus each $n_iX^i/g \in R(X)$. We have $tf - h = z/f_1$ where z is nilpotent in $R[X]$ and $f_1 \in R[X]$ such that $c(f_1) = R$. Then $tff_1 = hf_1 + z \in T(R)[X]$ and is integral over $R[X]$; hence $tff_1 \in R[X]$. Therefore $t \in R[X]_S = R(X)$.

Corollary 16.7 If R is an integrally closed Noetherian ring, then $R(X)$ is an integrally closed ring.

The proof of Theorem 16.6 shows that the conclusion of Theorem 16.3 holds for integrally closed Noetherian rings with respect to their total quotient rings—that is, if R is such a ring, then $R'(X)$ is the integral closure of $R(X)$ in $T(R)(X)$. Also note that unlike the ring $R[X], R(X)$ may be integrally closed even if the nilradical of R is nonzero. The next theorem gives another instance when this happens.

Theorem 16.8 If R is a Prüfer ring with Property A, then $R(X)$ is integrally closed.

PROOF: This proof proceeds along the same lines as the proof of Theorem 16.6. By the proof of Theorem 16.6, we need only show that if $t = nX^i/g$ where $g = a_0 + a_1X + \cdots + a_mX^m$ is a regular element of $R[X]$ and $n \in N(R)$, then $t \in R(X)$. By Property A, $c(g)$ is a regular ideal of R; hence it is invertible. Choose $b_0/s, \ldots, b_m/s \in c(g)^{-1}$ such that $\sum a_ib_i/s = 1$ where $b_i, s \in R$ and s is not a zero divisor. Let $h = b_0X^m + \cdots + b_m$. Since $a_ib_j/s \in R$ for all i, j and since the coefficient of X^m in gh is s, $f = gh/s \in R[X]$ and $c(f) = R$. The ring R is integrally closed, thus it contains all the nilpotent elements of $T(R)$. Therefore $t = nX^i/g = (nX^ih)/gh = (s^{-1}nX^ih)/f \in R(X)$.

A stronger version of Theorem 16.8 appears as Corollary 18.11. Theorem 16.6 yields an easier proof of Theorem 13.11, which we restate as a corollary.

Corollary 16.9 (Same as Theorem 13.11). If R is an integrally closed reduced ring satisfying Property A, then $R[X]$ is integrally closed.

PROOF: By Theorem 16.6, $R(X) = R(X)'$. If $t \in R[X]' \subseteq R(X)$, then there exists some $h \in R[X]$ such that $(t-h)^n = 0$ (Lemma 16.2). Therefore $t = h \in R[X]$.

We end this section with a sharper version of Theorems 13.3 and 13.11. There is a reduced integrally closed ring R such that $R[X]$ is integrally closed, but R does not have Property A, Example 15 of Section 27. Thus the converse of Theorem 13.11 fails. It is an open question as to whether or not the sufficient condition given in Theorem 16.10 is necessary.

Theorem 16.10 Let R be a reduced ring. If for each maximal ideal M of R, R_M is integrally closed with Property A, then $R[X]$ is integrally closed.

PROOF: For each $M \in \operatorname{Max} R$, $R_M[X]$ is an integrally closed ring, Theorem 13.11. Hence $R_M(X) = R(X)_{MR(X)}$ is integrally closed. By Lemma 13.2 and Theorem 14.1, $R(X)$ is integrally closed. Lemma 16.2 completes the proof.

17. The Ring $R\langle X\rangle$

Throughout this section denote the saturation of a multiplicatively closed set $\mathcal{U}$ by $\mathcal{U}^*$.

For a ring R, the set W of monic polynomials in $R[X]$ forms a regular multiplicatively closed subset of $R[X]$. Define $R\langle X\rangle$ to be the ring $R[X]_W$. Obviously $W \subseteq S$ where S is the set of $f \in R[X]$ with unit content. Hence $R(X)$ is a quotient ring of $R\langle X\rangle$. First the dimension of $R\langle X\rangle$ is determined in terms of $\dim R[X]$. Unlike the set S, which is saturated, W is usually not saturated. We describe the saturation of W in Theorem 17.10, and then give several applications of this result.

Lemma 17.1 Assume that R is a ring.

(1) If $\mathcal{M}$ is a maximal ideal of $R\langle X\rangle, \mathcal{Q} = \mathcal{M} \cap R[X]$ and $P = \mathcal{Q} \cap R$, then $R\langle X\rangle_{\mathcal{M}} = R[X]_{\mathcal{Q}} = R_P[X]_{\mathcal{Q}_{R\setminus P}}$.

(2) If N is a multiplicatively closed subset of R, then $R_N\langle X\rangle = R[X]_T$, where T is the set of $f \in R[X]$ whose leading coefficient is in N.

PROOF: (1): Since $\mathcal{M}$ is the extension of $\mathcal{Q}$, $R\langle X\rangle_{\mathcal{M}} = (R[X]_W)_{\mathcal{Q}_W} = R[X]_{\mathcal{Q}} = R_P[X]_{\mathcal{Q}_{R\setminus P}}$.

(2): This is routine.

Our first objective is to determine the $\dim R\langle X\rangle$.

Lemma 17.2 Let R be a finite dimensional ring. If $\mathcal{M}$ is a maximal ideal of $R[X]$ of maximal height, then $M = \mathcal{M} \cap R$ is a maximal ideal of R.

PROOF: By [G, p. 368] and [K, p. 25] we may find a chain of maximal length in $R[X]$ of the form

$$P \subset \cdots \subset M[X] \subset \mathcal{M} \tag{17a}$$

If M is not a maximal ideal in R, choose a maximal ideal M_1 of R properly containing M. Then $M[X] \subset M_1[X] \subset \mathcal{Q}$ for some maximal ideal $\mathcal{Q}$ of $R[X]$. This contradicts the maximality of (17a).

Theorem 17.3 If R is a finite dimensional ring, then $\dim R\langle X\rangle = \dim R[X] - 1$.

PROOF: Choose a maximal ideal $\mathcal{M}$ of $R[X]$ of maximal height. Then $M = \mathcal{M} \cap R$ is a maximal ideal of R. If $f \in \mathcal{M} \setminus M[X]$, then f can be modified so that its leading coefficient is not in M. Write $f = a_nX^n + \cdots + a_0$. Then $(M, a_n) = R$. For some $y \in R$ and $m \in M$, $m + a_ny = 1$. So $mX^n + yf$ is a monic polynomial belonging to $\mathcal{M}$. Since $\mathcal{M} \cap W \neq \varnothing$, since $M[X] \cap W = \varnothing$, and since there are no prime ideals properly between $\mathcal{M}$ and $M[X]$, we have $\dim R[X]_W = \dim R\langle X\rangle = \dim R[X] - 1$.

Corollary 17.4 If R is a finite dimensional ring, then $\dim R(X) = \dim R\langle X\rangle$.

PROOF: The maximal ideals of $R(X)$ are of the form $MR(X)$, where $M \in \operatorname{Max} R$.

Corollary 17.5 If R is a Noetherian ring of finite dimension, then $\dim R(X) = \dim R\langle X\rangle = \dim R$.

PROOF: By [G, p. 356], $\dim R[X] = \dim R + 1$.

For the ring $R(X) = R[X]_S$ the set S saturated, $S = R[X] \setminus \cup M[X]$ where M varies over $\operatorname{Max} R$, and the group of units of $R(X)$ is the set of elements f/g where f and g are in $R[X]$ and $c(f) = c(g) = R$. This situation is less clear for $R\langle X\rangle$. To find the group of units of $R\langle X\rangle$ we need to find the saturation W^* of W. To facilitate our work we introduce another multiplicatively closed set. Define $\mathcal{S}$ to be the set of polynomials in $R[X]$ with leading coefficients a unit. If I is an ideal of R, let $\mathcal{S}_I$ denote the polynomials in $(R/I)[X]$ with leading coefficients a unit. For an ideal I of R, let φ_I be the canonical homomorphism from R onto R/I and let $\varphi_I^* : R[X] \to (R/I)[X]$ be the homomorphism defined by $\varphi_I^*(a_0+a_1X+\cdots+a_nX^n) = \varphi_I(a_0)+\varphi_I(a_1)X+\cdots+\varphi_I(a_n)X^n$. Clearly $\varphi_I^*(\mathcal{S}) \subseteq \mathcal{S}_I$. Several preliminaries are needed before the characterization of W^* can be given. Note that $W^* = \mathcal{S}^*$. Most of our results are stated for $\mathcal{S}^*$.

Lemma 17.6 Assume that $f = a_0 + a_1X + \cdots + a_nX^n$ is in $\mathcal{S}^*$. Then:

(1) $c(f) = R$.

(2) For each a_j and for each prime ideal P of R, the relations $a_{j+1}, \dots, a_n \in P$, $a_j \notin P$ imply that a_j is a unit modulo P.

PROOF: (1): Since $\mathcal{S} \subseteq S$ and since S is saturated, $\mathcal{S}^* \subseteq S$.

(2): Fix a prime ideal P of R and assume that $a_{j+1}, \dots, a_n \in P$ and $a_j \notin P$. The saturated set $\mathcal{S}^*$ has the property that each element of $\mathcal{S}^*$ divides some element of $\mathcal{S}$ [G, p. 15]. Write $ff_1 = g \in \mathcal{S}$. Then $\varphi_P^*(f)\varphi_P^*(f_1) = \varphi_P^*(g) \in \mathcal{S}_P$. Since R/P is an integral domain, $\mathcal{S}_P$ is saturated. Therefore $\varphi_P^*(f) \in \mathcal{S}_P$ and hence the coset $a_j + P$ is a unit in R/P.

Polynomials satisfying (1) and (2) of Lemma 17.6 are called $(*)$-*polynomials.*

Lemma 17.7 Let N be the nilradical of a ring R and let $f \in R[X]$. Then f is a $(*)$-polynomial if and only if $\varphi_N^*(f)$ is a $(*)$-polynomial.

PROOF: Let $f = a_0 + a_1X + \cdots + a_nX^n$. Since $(a_0, \dots, a_n) = R$, if and only if $(\varphi_N(a_0), \dots, \varphi_N(a_n)) = R/N$, condition (1) of Lemma 17.6 holds for f if and only if it holds for $\varphi_N^*(f)$.

Consider condition (2) of Lemma 17.6. There is a one-to-one correspondence between the prime ideals of R and those of R/N given by $P \leftrightarrow P/N$. Let P be a prime ideal of R and let j be an integer between 0 and n such that $a_{j+1}, \dots, a_n \in P$ and $a_j \notin P$. Then $\varphi_N(a_{j+1}), \dots, \varphi_N(a_n) \in P/N$ and $\varphi_N(a_j) \notin P/N$. Note that $R \xrightarrow{\varphi_N} R/N \xrightarrow{\varphi_{P/N}} (R/N)/(P/N) = R/P$ and $\varphi_P = \varphi_{P/N}\varphi_N$. Thus a_j is a unit modulo P if and only if $\varphi_N(a_j)$ is a unit modulo P/N.

Lemma 17.8 If $f \in R[X]$, then $f \in \mathcal{S}^*$ if and only if $\varphi_N^*(f) \in \mathcal{S}_N^*$.

PROOF: ($\Rightarrow$): If f divides a monic polynomial g of $R[X]$, then $\varphi_N^*(f)$ divides the monic polynomial $\varphi_N^*(g)$ of $(R/N)[X]$. Hence $\varphi_N^*(f) \in \mathcal{S}_N^*$.

($\Leftarrow$): If $\varphi_N^*(f) \in \mathcal{S}_N$, then there exists a monic polynomial $\varphi_N^*(g) \in (R/N)[X]$ such that $\varphi_N^*(f)\varphi_N^*(h) = \varphi_N^*(g)$, for some $h \in R[X]$. Thus $fh - g \in N[X] = \ker \varphi_N^*$. Therefore $(fh-g)^k = 0$ for some k. It follows that f divides the monic polynomial g^k, hence $f \in \mathcal{S}^*$.

Lemma 17.9 Let b be an element of a reduced ring R. If b is a unit modulo P for each P not containing b, then (b) is an idempotent ideal of R.

PROOF: It suffices to prove that (b) is locally idempotent. We need only consider the case where M is a maximal ideal of R containing b. If P is a prime ideal contained in M, then $\varphi_P(b)$ is not a unit in R/P, hence b belongs to each prime ideal contained in M. Therefore $bR_M \subseteq N(R_M) = (0)_M$.

Theorem 17.10 Assume the preceding notation. Then W^* is the set of $(*)$-polynomials over R.

PROOF: It suffices to show that $\mathcal{S}^*$ is the set of $(*)$-polynomials of R. By Lemma 17.6, each element of $\mathcal{S}^*$ is a $*$-polynomial. By Lemmas 17.7 and 17.8 it is sufficient to establish the opposite inclusion when R is assumed to be a reduced ring.

Let $f = d_0 + d_1X + \cdots + d_nX^n$ be a $*$-polynomial over R. The proof is by induction on n. For $n = 0$, $c(f) = (d_0) = R$. Hence $f \in \mathcal{S}^*$. Assume that the result has been established for $(*)$-polynomials of degree $< n$ and let f be a $(*)$-polynomial of degree $= n$. The leading coefficient d_n of f generates an idempotent ideal of R, Lemma 17.9. If d_n is a unit in R, then $f \in \mathcal{S} \subseteq \mathcal{S}^*$ and we are finished. Assume that $(d_n) = A \subset R$. Decompose R into the direct sum $A \oplus B$ where $B = \mathrm{Ann}(d_n)$. For each coefficient d_i of f, write $d_i = a_i + b_i$ where $a_i \in A$ and $b_i \in B$.

Since $R[X] = A[X] + B[X]$ we have $f = g + h$ where $g = a_0 + a_1X + \cdots + a_nX^n \in A[X]$ and $h = b_0 + b_1X + \cdots + b_nX^n \in B[X]$. Since A and B are ideals of R and since $(A \oplus B)/B = A$, $\mathcal{S}_B$ is the set of polynomials in $A[X]$ whose leading coefficients are units. Similarly for $\mathcal{S}_A$. It is straightforward to see that if $g \in \mathcal{S}_B^*$ and $h \in \mathcal{S}_A^*$, then $g + h \in \mathcal{S}^*$. With this in mind we show that $g \in \mathcal{S}_B^*$ and $h \in \mathcal{S}_A^*$. Note that $d_n = a_n$ and $b_n = 0$, so $\deg h = m < n$. Since d_n is a unit of A, $g \in \mathcal{S}_B \subseteq \mathcal{S}_B^*$. Certainly $c_R(f) \subseteq c_A(g) \oplus c_B(h)$. But $c_R(f) = R$ implies that $c_R(f) = c_A(g) \oplus c_B(h) = R$. Thus $c_B(h) = B$. Condition (1) of Lemma 17.6 is satisfied by h with respect to the ring $B[X]$. We claim that condition (2) of Lemma 17.6 holds for h and $B[X]$. Assume that j is an integer between 0 and m, and P is a prime ideal of B such that $b_{j+1}, \ldots, b_m \in P$, $b_j \notin P$. Then $A \oplus P$ is a prime ideal of R, $d_{j+1}, \ldots, d_n \in A \oplus P$, and $d_j \notin A \oplus P$. By hypothesis $d_j = a_j + b_j$ is a unit modulo $A \oplus P$. This implies that b_j is a unit modulo P. We have proved that h is a $(*)$-polynomial in $B[X]$. Since $\deg h < n$, the induction hypothesis implies that $h \in \mathcal{S}_A^*$.

Two applications of Theorem 17.10 follow.

Theorem 17.11 The rings $R(X)$ and $R\langle X\rangle$ coincide if and only if R is zero-dimensional.

PROOF: Note that $R\langle X\rangle = R(X)$ if and only if $S = W^*$.

($\Leftarrow$): For $f \in S$ we must show that f is a $(*)$-polynomial. Without loss of generality assume that R is reduced. Hence R is a von Neumann regular ring (Remark following Corollary 2.6). By definition, $c(f) = R$. A finitely generated ideal of R is idempotent (ibid.). Write $f = a_0 + a_1X + \cdots + a_nX^n$. If P is a prime ideal of R and if j is an integer such that $a_{j+1}, \ldots, a_n \in P$ and $a_j \notin P$, then as ideals $(\varphi_P(a_j)) = (\varphi_P(a_j))^2$ in the integral domain R/P. Thus $\varphi_P(a_j)$ is a unit in R/P.

($\Rightarrow$): Assume $\dim R > 0$ and let $P \subset M$ be proper prime ideals of R. Let $m \in M \setminus P$ and define $f = 1 + mX$. Then $f \in S$, but f is not a $(*)$-polynomial since m is not a unit modulo P. Therefore $S \neq W^*$, a contradiction.

For two indeterminates X and Y, define $R\langle X,Y\rangle = R\langle X\rangle\langle Y\rangle$. A major difference between $R\langle X,Y\rangle$ and $R(X,Y)$ now emerges. (Compare Theorem 17.12 with Lemma 15.3.)

Theorem 17.12 Let X and Y be indeterminates over R. Then $R\langle X,Y\rangle = R\langle Y,X\rangle$ if and only if $\dim R = 0$.

PROOF: ($\Rightarrow$): Assume that R contains proper ideals $P \subset M$. Choose $m \in M \setminus P$ and let $f = X + mY$. Then f is a monic polynomial with respect to X in $R\langle Y\rangle[X]$. Thus f is a unit in $R\langle Y\rangle\langle X\rangle$. We claim that f is not a unit in $R\langle X\rangle\langle Y\rangle$. (This will show that $R\langle Y,X\rangle \neq R\langle X,Y\rangle$.) As before it is sufficient to prove this claim under the assumption that $R\langle X\rangle$ is a reduced ring. Suppose that f is actually a unit in $R\langle X\rangle\langle Y\rangle$. Then f is a $(*)$-polynomial of $R\langle X\rangle[Y]$. Hence m generates an idempotent ideal of $R\langle X\rangle$, Lemma 17.9. Since $PR\langle X\rangle$ is a prime ideal of $R\langle X\rangle$, $\varphi_{PR\langle X\rangle}(m)$ generates an idempotent ideal of the domain $R\langle X\rangle/PR\langle X\rangle$. There are two cases: (1) If $m \notin PR\langle X\rangle$, then $\varphi_{PR\langle X\rangle}(m) \in MR\langle X\rangle/PR\langle X\rangle$ which is impossible; (2) If $m \in PR\langle X\rangle$, then $m \in PR\langle X\rangle \cap R = P$, a contradiction. Therefore f is not a unit in $R\langle X\rangle\langle Y\rangle$.

($\Leftarrow$): By Theorem 17.3 $\dim R\langle X\rangle = \dim R[X] - 1$ and since $\dim R[X] = 1$ [K, p. 26], $\dim R\langle X\rangle = 0$. Complete the proof by using Theorem 17.11.

18. Divisibility in $R\langle X\rangle$ and $R(X)$

In this section a comparison of $R\langle X\rangle$ and $R(X)$ is made, mostly with respect to divisibility properties. Several of the results on divisibility must first be established for integral domains. We consistently use D for an integral domain and R for an arbitrary commutative ring. When it is possible to state the results just for rings we do so.

The proof of the first result is about the same as the corresponding proof for integral domains.

Lemma 18.1 Let I be an ideal of a ring R. Then I is a finitely generated, regular, locally principal ideal of R if and only if I is invertible.

PROOF: ($\Rightarrow$): If M is a maximal ideal of R, then I_M is a regular principal ideal in R_M, hence $I_M(R_M : I_M) = R_M$. Since I is finitely generated, $(R_M : I_M) = (R : I)_M$. Consequently $I_M(R_M : I_M) = I_M(R : I)_M = R_M$ for each maximal ideal M of R. Therefore $I(R : I) = R$.

($\Leftarrow$): It is obvious that I is a regular finitely generated ideal of R. Write $1 = \sum_{i=1}^{n} a_i b_i$ where $a_i \in I$ and $b_i \in I^{-1}$. Writing $a/1$ for the image of a in R_M we have $1/1 = \sum (a_i/1)(b_i/1)$. Say that $(a_1/1)(b_1/1) \notin M_M$. Thus $(a_1/1)(b_1/1)$ is a unit in R_M and hence $(a_1/1)R_M = I_M$.

Theorem 18.2 Let I be an ideal of a ring R.

(1) $IR\langle X\rangle \cap R = I$.
(2) The following are equivalent:
  (a) I is finitely generated and locally principal.
  (b) $IR\langle X\rangle$ is finitely generated and locally principal.
  (c) $IR(X)$ is principal.
(3) The following are equivalent:
  (a) $IR(X)$ is invertible.
  (b) $IR\langle X\rangle$ is invertible.
  (c) I is dense, finitely generated, and locally principal.

PROOF: The proof of (1) is clear.

(2): Assume that I is finitely generated and locally principal. Let $\mathcal{M}$ be a maximal ideal of $R\langle X\rangle$, $\mathcal{Q} = \mathcal{M} \cap R[X]$, and $P = \mathcal{Q} \cap R$. Then $R\langle X\rangle_{\mathcal{M}} = R_P[X]_{\mathcal{Q}_{R\setminus P}}$, Lemma 17.1. Since I_P is principal, so is $IR\langle X\rangle_{\mathcal{M}} = I_P R_P[X]_{\mathcal{Q}_{R\setminus P}}$. Since $R(X)$ is a quotient ring of $R\langle X\rangle$, $IR(X)$ is finitely generated and locally principal. By Corollary 15.2, $IR(X)$ is principal. Another application of Corollary 15.2 completes the proof of (2).

(3): Assume that $IR\langle X\rangle$ is invertible. Since $R(X)$ is a quotient ring of $R\langle X\rangle$, $IR(X)$ is also invertible. Thus $IR(X)$ is finitely generated and locally principal (Lemma 18.1). Therefore I is a finitely generated locally principal ideal of R, Corollary 15.2. Let $I = (a_0, \ldots, a_n)$ and $f = a_0 + a_1X + \cdots + a_nX^n$. By Theorem 15.1, $IR(X) = fR(X)$. Since invertible ideals are regular, f is regular. Hence I is a dense ideal.

We still need to show that $(c) \Rightarrow (b)$. Assume that $I = (a_0, \ldots, a_n)$ is locally principal and dense. Then $\mathcal{I} = IR\langle X\rangle$ is finitely generated and locally principal. The element $f = a_0 + a_1X + \cdots + a_nX^n$ is regular in $\mathcal{I}$. To complete the proof apply Lemma 18.1.

Use GCD-domain to mean greatest common divisor domain and UFD to mean unique factorization domain. Write $\gcd\{a,b\}$ for the greatest common divisor of a and b.

Theorem 18.3 If D is an integral domain, then:

(1) *D is a GCD-domain if and only if $D\langle X\rangle$ is a GCD-domain.*

(2) *D is a UFD if and only if $D\langle X\rangle$ is a UFD.*

PROOF: (1) ($\Rightarrow$): The GCD-property is preserved under adjunction of an indeterminate and quotient ring formation [G, pp. 424 and 182].

(1) ($\Leftarrow$): Assume that $a, b \in D$. Then $\gcd\{a,b\}$ exists in $D\langle X\rangle$; say that $f/h = \gcd\{a,b\}$ where $f, h \in D[X]$ and h is

monic. If c is the leading coefficient of f, then c divides a and b in D. If $d \in D \subset D\langle X\rangle$ such that d divides a and b, then d divides f/h (in $D\langle X\rangle$). An easy argument shows that d divides the leading coefficient c of f. Therefore $c = \gcd\{a, b\}$.

(2) ($\Rightarrow$): This is clear.

(2) ($\Leftarrow$): Suppose that $D\langle X\rangle$ is a UFD. Then $D\langle X\rangle$ is a GCD-domain and hence D is a GCD-domain. By [G, p. 176], a GCD-domain in which every ascending chain of principal ideals is finite is a UFD. Let $(a_1) \subseteq (a_2) \subseteq \cdots$ be an ascending chain of principal ideals in D. For some i, $a_{i-1}D\langle X\rangle = a_iD\langle X\rangle$. Then $a_i = a_{i-1}f/g$ where $f, g \in D[X]$ and g is monic. Then $a_i = ca_{i-1}$ where c is the leading coefficient of f. Therefore $(a_{i-1}) = (a_i)$.

Theorem 18.4 For an integral domain D, the following are equivalent:

(1) D is Krull domain.
(2) $D\langle X\rangle$ is a Krull domain.
(3) $D(X)$ is a Krull domain.

PROOF: (1) $\Rightarrow$ (2) and (2) $\Rightarrow$ (3) follow from properties of adjunction of an indeterminate and of quotient rings [G, pp. 528 and 530].

(3) $\Rightarrow$ (1): Let K be the quotient field of D. Then $D(X) \cap K = D$ is a Krull domain.

Theorem 18.5 An integral domain D is a principal ideal domain (PID) if and only if $D\langle X\rangle$ is a principal ideal domain.

PROOF: Note that D is a field if and only if $D\langle X\rangle$ is a field, Theorem 17.3. Assume that D is a one-dimensional PID; then $D\langle X\rangle$ is a one-dimensional integrally closed Noetherian domain. Therefore $D\langle X\rangle$ is a Dedekind domain [G, p. 446] that is also a UFD (Theorem 18.3). Thus $D\langle X\rangle$ is a PID [G, p. 464]. Conversely, assume that $D\langle X\rangle$ is a PID. By Theorems 18.3 and 18.4, D is a Krull domain which is also a UFD. Then each height one prime ideal of D is principal [G, pp. 534–535]. Since D is one-dimensional, D is a PID [G, p. 447].

Theorem 18.6 For an integral domain D the following conditions are equivalent:

(1) D is a Dedekind domain.
(2) $D\langle X\rangle$ is a Dedekind domain.
(3) $D(X)$ is a Dedekind domain.
(4) $D(X)$ is a PID.

PROOF: (1) $\Rightarrow$ (2): Since $\dim D = \dim D\langle X\rangle$ (Corollary 17.5), $D\langle X\rangle$ is a one-dimensional integrally closed Noetherian domain—that is, a Dedekind domain.

(2) $\Rightarrow$ (3): Overrings of Dedekind domains are Dedekind domains.

(3) $\Rightarrow$ (4): Theorem 15.4.

(4) $\Rightarrow$ (1): From [G, p. 413], D is a Prüfer domain. It remains to show that D is a Noetherian domain. If I is an ideal of D, then $ID(X)$ is a principal ideal. Choose $fD[X]$ such that the coefficients of f are in I and so that $ID(X) = fD(X)$. Write $f = a_0 + a_1X + \cdots + a_nX^n$ where $a_i \in I$. By Theorem 15.1, $c(f)D(X) = fD(X) = ID(X)$. By Theorem 14.1(3), $(a_0, \ldots, a_n) = c(f)D(X) \cap D = ID(X) \cap D = I$.

We generalize to the zero divisor case. Let R be a finite direct sum of subrings $R = R_1 \oplus \cdots \oplus R_n$. Then $R(X) = R_1(X) \oplus \cdots \oplus R_n(X)$. On the other hand, assume that $R(X) = T_1 \oplus \cdots \oplus T_n$ where each T_i is a subring of $R(X)$. If e_i is the identity of T_i, then $1 = (e_1, \ldots, e_n)$. Each e_i is idempotent and belongs to $R(X)$. Hence each $e_i \in R$, Theorem 14.7. If $R_i = Re_i$, then $R = R_1 \oplus \cdots \oplus R_n$, $T_i = R(X)e_i = R_i(X)$, and $R(X) = R_1(X) \oplus \cdots \oplus R_n(X)$.

The above paragraph holds if $R(X)$ is replaced by either $R[X]$ or $R\langle X\rangle$.

A principal ideal ring (PIR) is *special* (SPIR) if R has exactly one prime ideal $P \neq R$ and P is nilpotent. If every ideal of R is a (finite) product of prime ideals, then R is called a general ZPI-*ring*. (A ZPI-*ring* is one in which the factorization is unique.) By [G, p. 470], R is a general ZPI-ring if and only if R is a finite direct sum of Dedekind domains and SPIRs.

Lemma 18.7 For a ring R the following conditions are equivalent:

(1) R is an SPIR.
(2) $R\langle X\rangle$ is an SPIR.
(3) $R(X)$ is an SPIR.

PROOF: (1) $\Leftrightarrow$ (3): If R is an SPIR with unique maximal ideal M, then $MR(X)$ is the unique maximal ideal of $R(X)$, Theorem 14.1, and $M^n = (0)$ implies that $MR(X)^n = (0)$. Let $A = (f_1, \ldots, f_t)$, where the $f_i \in R[X]$, be an arbitrary ideal of $R(X)$. By Theorem 15.1, $c(f_i)R(X) = f_iR(X)$. If $(c(f_1), \ldots, c(f_t)) = (b)$, then $bR(X) = (f_1, \ldots, f_t)R(X) = A$. Therefore $R(X)$ is an SPIR. Conversely, if $R(X)$ is an SPIR, let $MR(X)$ be its unique prime ideal. Then M is the unique prime ideal of R. Since $MR(X)$ is nilpotent, so is M.

(1) $\Leftrightarrow$ (2): If R is an SPIR, then $R(X) = R\langle X\rangle$, Theorem 17.11. Conversely, assume that $R\langle X\rangle$ is an SPIR.Then $\dim R\langle X\rangle = \dim R(X)$; and by the equivalence of (1) and (3), R is an SPIR.

Theorem 18.8 If R is a ring, then the following conditions are equivalent:

(1) R is a general ZPI-ring.
(2) $R\langle X\rangle$ is a general ZPI-ring.
(3) $R(X)$ is a general ZPI-ring.
(4) $R(X)$ is a PIR.

PROOF: (1) $\Rightarrow$ (2): Write $R = R_1 \oplus \cdots \oplus R_n$ where R_i is either a Dedekind domain or an SPIR.Then $R\langle X\rangle = R_1\langle X\rangle \oplus \cdots \oplus R_n\langle X\rangle$. By the preceding two results, each $R_i\langle X\rangle$ is a Dedekind domain or an SPIR.

(2) $\Rightarrow$ (3): If $R\langle X\rangle$ is a general ZPI-ring, then $R\langle X\rangle = R_1\langle X\rangle \oplus \cdots \oplus R_n\langle X\rangle$ where each $R_i\langle X\rangle$ is a Dedekind domain or an SPIR.Thus each R_i is Dedekind or an SPIR.We conclude that

$$R(X) = R_1(X) \oplus \cdots \oplus R_n(X) \tag{18a}$$

is a general ZIP-ring.

(3) $\Rightarrow$ (4): Consider the representation (18a) of $R(X)$. If R_i is a Dedekind domain, then $R_i(X)$ is a PID. Thus each $R_i(X)$ is a PIR. But a finite sum of PIRs is still a PIR.

(4) $\Rightarrow$ (1): If $R(X)$ is a PIR, then the factors of $R(X)$ in (18a) are PIRs. If $R_i(X)$ is a domain, then R_i is a Dedekind domain. If $R_i(X)$ is an SPIR, then so is R_i. Therefore R is a general ZPI-ring.

Theorem 18.9 The ring R is a PIR if and only if $R\langle X\rangle$ is a PIR.

PROOF: A ring R is a PIR if and only if R is a finite direct sum of PIDs and SPIRs [G, p. 475]. If R is a PIR, then Theorem 18.5 and Lemma 18.7 imply that $R\langle X\rangle$ is a PIR.

Conversely assume that $R\langle X\rangle$ is a PIR. Then $R\langle X\rangle = R_1\langle X\rangle \oplus \cdots \oplus R_n\langle X\rangle$ where $R = R_1 \oplus \cdots \oplus R_n$. Since $R_i\langle X\rangle$ is a PID or an SPIR, R is a PIR.

Next we determine when $R(X)$ and $R\langle X\rangle$ are Prüfer rings. A ring R is *strongly Prüfer* if every finitely generated dense ideal is locally principal. Let I be a finitely generated dense ideal in a Prüfer ring with Property A. Then I is regular and is therefore invertible. Consequently I is locally principal, Lemma 18.1. Thus a Prüfer ring with Property A is strongly Prüfer. That a strongly Prüfer ring is a Prüfer ring is also a consequence of Lemma 18.1. Hence: Prüfer + Property A $\Rightarrow$ strongly Prüfer $\Rightarrow$ Prüfer.

Theorem 18.10 Let R be a ring. Then $R(X)$ is a Prüfer ring if and only if R is a strongly Prüfer ring.

PROOF: ($\Rightarrow$): Let I be a finitely generated dense ideal of R. If $I = (a_0, \ldots, a_n)$, then $f = a_0 + a_1X + \cdots + a_nX^n$ is a regular element belonging to $IR(X)$. Thus $IR(X)$ is an invertible ideal of the Prüfer ring $R(X)$. By Lemma 18.1, $IR(X)$ is locally principal. Therefore I is a locally principal ideal (Corollary 15.2).

($\Leftarrow$): Let $\mathcal{I}$ be a finitely generated regular ideal of $R(X)$. We choose $f_0, \ldots, f_n \in R[X]$ such that $\mathcal{I} = (f_0, \ldots, f_n)R(X)$. Let $t = \max\{\deg f_i\}$ and define $g = f_0 + X^{t+1}f_1 + \cdots + X^{n(t+1)}f_n$. Then $c(g)$ is a finitely generated dense ideal of R and is therefore locally principal. By Theorem 15.1, $gR(X) = c(g)R(X)$, which implies that $\mathcal{I} = gR(X)$.

Let P be a prime ideal in R and let $a(X)$ be a monic irreducible polynomial in $T(R/P)[X]$. Define $\langle P, a(X)\rangle = \{g \in R[X] : a(X) \text{ divides } \bar{g}\}$, where $\bar{g}$ is the polynomial obtained from g by reducing its coefficient modulo P. If $\mathcal{M}$ is a prime ideal in $R[X]$ such that $\mathcal{M} \cap R = P$, then either $P[X] = \mathcal{M}$ or $\mathcal{M} \supset P[X]$. With the help of [K, p. 25] it is easy to see in the latter case that $\mathcal{M} = \langle P, a(X)\rangle$ for some $a(X)$.

With the aid of Theorem 18.10 we are able to strengthen Theorem 16.8 to the following.

Corollary 18.11 [cf. Theorem 16.8] If R is a Prüfer ring with Property A, then $R(X)$ is a Prüfer ring. Hence $R(X)$ is integrally closed.

PROOF: Since R is a Prüfer ring with Property A, R is strongly Prüfer. Hence $R(X)$ is a Prüfer ring.

Theorem 18.12 For a ring R the following conditions are equivalent:

(1) $R\langle X\rangle$ is a Prüfer ring.

(2) R is strongly Prüfer, $\dim R \leq 1$, and if $P \subset Q$ are prime ideals of R, then R_P is a field.

PROOF: (2) $\Rightarrow$ (1): Let $\mathcal{K}$ be a finitely generated regular ideal of $R\langle X\rangle$. We prove that $\mathcal{K}$ is invertible by showing that if $\mathcal{M} \supseteq \mathcal{K}$ is a maximal ideal of $R\langle X\rangle$, then $\mathcal{K}_{\mathcal{M}}$ is principal. Write $\mathcal{K} = \mathcal{I}_W$ where $\mathcal{I}$ is a finitely generated regular ideal of $R[X]$, and $\mathcal{M} = \mathcal{Q}_W$ where $\mathcal{Q}$ is a prime ideal $R[X]$. Let $P = \mathcal{Q} \cap R$. We show that $\mathcal{K}_{\mathcal{M}} = \mathcal{I}_{\mathcal{Q}}$ is principal. If P is not a maximal ideal of R, then R_P is a field. Hence $R[X]_{\mathcal{Q}} \cong R_P[X]_{\mathcal{Q}_{R\setminus P}}$ is a valuation domain and thus $\mathcal{I}_{\mathcal{Q}}$ is a principal ideal.

Now assume that P is a maximal ideal. Then either $\mathcal{Q} = P[X]$ or $\mathcal{Q} = \langle P, a(X)\rangle$. Suppose that $\mathcal{Q} = \langle P, a(X)\rangle$. Since P is maximal, $T(R/P) = R/P$, so $a(X) = X^n + \bar{a}_{n-1}X^{n-1} + \cdots + \bar{a}_0$ where $\bar{a}_i$ is the residue of a_i modulo P, $a_i \in R$. Then $f = X^n + a_{n-1}X^{n-1} + \cdots + a_0 \in \mathcal{Q}$. This is impossible since $\mathcal{Q} \cap W = \emptyset$. Therefore $\mathcal{Q} = P[X]$, and hence $R[X]_{\mathcal{Q}} = R[X]_{P[X]} = R_P(X)$. Let $\mathcal{I} = (f_0, \ldots, f_n)$, $t = \max\{\deg f_i\}$, and $g = f_0 + X^{t+1}f_1 + \cdots + X^{n(t+1)}f_n$. As in the proof of Theorem 18.10, $\mathcal{I}_{\mathcal{Q}} = \mathcal{I}R_P(X) = gR_P(X)$.

(1) $\Rightarrow$ (2): Since $R\langle X\rangle$ is a Prüfer ring, the overring $R(X)$ is also Prüfer. Thus R must be a strongly Prüfer ring (Theorem 18.10). Let $P_0 \subset P$ be prime ideals in R. Choose $p \in P_0$ and $m \in P \setminus P_0$. Then $\mathcal{I} = (p, mX+1)$ is a finitely generated regular ideal of $R[X]$. Hence $\mathcal{I}R\langle X\rangle$ is invertible. By looking at the leading coefficients of polynomials that can appear in $(P_0[X], mX+1)$ we see that $(P_0[X], mX+1) \cap W = \emptyset$. Thus there exists some prime ideal $\mathcal{Q}$ of $R[X]$ containing $(P_0[X], mX+1)$ such that $\mathcal{Q} \cap W = \emptyset$, Therefore $\mathcal{I}R[X]_{\mathcal{Q}} = \mathcal{I}R\langle X\rangle_{\mathcal{Q}R\langle X\rangle}$ is a principal ideal. Either p or $mX+1$ generates $\mathcal{I}R\langle X\rangle_{\mathcal{Q}}$. If p is the generator, then

$$\frac{mX+1}{1} = \frac{pf}{g}$$

where f and g are in $R[X]$, $g \notin Q$. There is an $s \in R[X] \setminus Q$ such that $sg(mX+1) = sfp \in P_0[X]$. This implies that $mX + 1 \in P_0[X]$, a contradiction. Therefore $\mathcal{I}R[X]_Q = (mX+1)R[X]_Q$; and hence $pR[X]_Q \subseteq (mX+1)R[X]_Q$. Write

$$\frac{p}{1} = \frac{(mX+1)f}{g}$$

where $f, g \in R[X]$ and $g \notin Q$. Then $pgh = (mX+1)fh$ for some $h \in R[X] \setminus Q$. Since the content of $mX+1$ is R, $c(fh) = c((mX+1)fh) = c(pgh) \subseteq (p)$. Then $fh = pt$ for an appropriate $t \in R[X]$. Thus

$$\frac{p}{1} = \frac{(mX+1)fg}{gh} = \frac{p(mX+1)t}{gh}$$

Consequently $pR[X]_Q = (QR[X]_Q)(pR[X]_Q)$. By Nakayama's lemma $pR[X]_Q = (0)$. Therefore $P_0R[X]_Q = (0)$. This proves that (0) is a prime ideal in $R[X]_Q$ and that $\dim R \leq 1$. Thus $(R[X]_Q)_{P_0R[X]_Q} = R[X]_{P_0[X]} = R_{P_0}(X)$ is a field. Therefore R_{P_0} is a field.

With the introduction of the concept of strongly Prüfer, we are able to give a new sufficient condition for $R[X]$ to be integrally closed.

Theorem 18.13 Let R be a reduced integrally closed ring for which $T(R)$ is strongly Prüfer. Then $R[X]$ is an integrally closed ring.

PROOF: We claim that in a reduced ring R, $R[X]$ is integrally closed if and only if $R(X)$ is integrally closed. The necessity is clear. Assume that $R(X)$ is integrally closed and let f/g be integral over $R[X]$. By Theorem 16.2, there exists an element $h \in R[X]$ such that $(f/g - h)^n = 0$, that is, $f/g = h \in R[X]$.

Since $T(R)$ is a strongly Prüfer ring, $T(R)(X)$ is a Prüfer ring (Theorem 18.10); hence $T(R)(X)$ is integrally closed. By the first paragraph $T(R)[X]$ is integrally closed. It follows from Theorem 13.4 that $R[X]$ is integrally closed.

The converse of Theorem 18.13 fails; see Example 18 of Section 27.

19. Multiplicatively Closed Subsets of $R[X]$

We introduce several new multiplicatively closed subsets of $R[X]$. The new rings derived from these sets will be used later to characterize Prüfer v-multiplication rings. Assume that $a, b \in R$ and write $(a : b)$ in place of $(a :_R b)$. We are interested in the following sets:

$$\mathcal{P}_1 = \{P \in \operatorname{Spec} R : P \text{ is minimal over } (a : b),\ a, b \in R\}$$

$$\mathcal{P}_2 = \{P \in \operatorname{Spec} R : P \text{ is minimal over } (a : b),\ a, b \in R \text{ with } a \text{ regular}\}$$

$$\mathcal{U}_1 = R[X] \setminus \bigcup_{P \in \mathcal{P}_1} P[X]$$

$$\mathcal{U}_2 = R[X] \setminus \bigcup_{P \in \mathcal{P}_2} P[X]$$

In the next proof we use two different localizations of R at P; R_P the usual localization and $R_{(P)}$ the regular localization (see Section 6). Thus when working in R_P, a/s may not be an element in $T(R)$, while in $R_{(P)}$ a/s is always in $T(R)$. Recall that if N is any multiplicatively closed subset of R, then $(N) = \{x \in N : x \text{ is regular}\}$. Clearly (N) is a multiplicatively closed set.

Theorem 19.1 If R is a ring, then:

(1) $\mathcal{U}_2 = \{f \in R[X] : c(f)^{-1} = R\}$.
(2) $R = \cap_{P\in\mathcal{P}_1} R_{(P)} = \cap_{P\in\mathcal{P}_2} R_{(P)}$.
(3) $\mathcal{U}_1$ is a regular multiplicatively closed set in $R[X]$.

PROOF: (1): If $f \in \mathcal{U}_2$, then $c(f) \not\subseteq P$ for all $P \in \mathcal{P}_2$. Let $a/b \in T(R)$ such that $(a/b)c(f) \subseteq R$. Thus $c(f) \subseteq (b : a) \subseteq R$, where b is a regular element in R. If $(b : a) \neq R$, then there is a $P \in \mathcal{P}_2$ such that $c(f) \subseteq (b : a) \subseteq P$. This contradiction implies that $(b : a) = R$; or equivalently, that $a/b \in R$. Hence $c(f)^{-1} = R$, and so $\mathcal{U}_2 \subseteq \{f \in R[X] : c(f)^{-1} = R\}$.

For the reverse inclusion, let $f \in R[X]$ such that $c(f)^{-1} = R$, and assume that $f \notin \mathcal{U}_2$. There exists a $P \in \mathcal{P}_2$ with $c(f) \subseteq P$. Assume that P is minimal over $J = (a : b)$, where a is a regular element in R. Localizing at P yields $P_P = \operatorname{Rad} J_P$. Let $c(f) = (a_0, \ldots, a_n) = I$. Thus in the ring R_P, $(a_i/1)^{t_i} = r_i/s_i$ where $r_i \in J$ and $s_i \in R \setminus P$. This implies that there exists a natural number t and some $s \in R\setminus P$ such that $sI^t \subseteq J$. Let k be the least nonnegative integer such that $sI^k \subseteq J$ and note that $k > 0$ since $s \notin J$. Then $(sb/a)I^k \subseteq R$, and hence $(sb/a)I^{k-1} \subseteq I^{-1} = R$, contradicting the minimality of k. Therefore $f \in \mathcal{U}_2$.

(2): It suffices to prove that $\cap_{P\in\mathcal{P}_2} R_{(P)} \subseteq R$. Let $b/a \in \cap_{P\in\mathcal{P}_2} R_{(P)}$, where $a, b \in R$ and a is a regular element. If $(a : b) \neq R$, choose some $P \in \mathcal{P}_2$ minimal over $(a : b)$. Then $b/a = r/s$ (as elements in $T(R)$) where $r, s \in R$, s regular and $s \in R \setminus P$. The equation $bs = ar$ implies that $s \in (a : b) \subseteq P$ which is impossible.

(3): If f is a zero divisor in $\mathcal{U}_1$, then there exists a nonzero $a \in R$ such that $af = 0$. Hence $c(f) \subseteq (0 : a)$. If P is a minimal prime ideal of $(0 : a)$, then $c(f) \subseteq P$. Therefore $f \in P[X]$, a contradiction.

Corollary 19.2 Let I be a finitely generated ideal of R. Then $I \not\subseteq P$ for all $P \in \mathcal{P}_2$ if and only if $I^{-1} = R$.

There are several things to note. The regular multiplicatively closed sets $\mathcal{U}_1$ and $(\mathcal{U}_2)$ determine overrings, $R[X]_{\mathcal{U}_1}$ and $R[X]_{(\mathcal{U}_2)}$, of $R[X]$. Of course the ring $R[X]_{\mathcal{U}_2}$ may not be an overring of $R[X]$. We have the following chain of multiplicatively closed sets:

$$W^* \subseteq S \subseteq \mathcal{U}_1 \subseteq (\mathcal{U}_2) \subseteq \mathcal{U}_2 \tag{19a}$$

In the integral domain case $\mathcal{U}_1 = (\mathcal{U}_2) = \mathcal{U}_2$. However in the arbitrary ring case all of these containments may be strict. We have already seen that $W^* \subset S$, unless $\dim R = 0$. Let I be a finitely generated ideal containing an R-sequence of length 2 in an integral domain D. Then $I^{-1} = D$ [K, p. 102]. If $I = (a_0, \ldots, a_n)$, then $f = a_0 + a_1 X + \cdots + a_n X^n \in D[X]$ is an element in $\mathcal{U}_1$ but not in S. Our next example shows that $\mathcal{U}_2$ and $(\mathcal{U}_2)$ may differ. Let $R = Z \oplus Q$ where Z is the ring of integers and Q is the field of rational numbers. If $f = (1,0) + (1,0)X \in R[X]$, then $c(f) = (1,0)R = M$ is a maximal ideal of R. Since f is a zero divisor it is not in $(\mathcal{U}_2)$. Suppose that $f \notin \mathcal{U}_2$. Then there exists a $P \in \mathcal{P}_2$ such that $f \in P[X]$. Hence $M = c(f) \subseteq P$, and by the maximality of M, $M = P \in \mathcal{P}_2$. This contradicts the fact that M consists entirely of zero divisors.

We need the following generalization of Exercise 1 of [K, p. 102].

Lemma 19.3 If an ideal I of a ring R contains an R-sequence of length 2, then $I^{-1} = R$.

PROOF: Suppose that $\{a, b\}$ is an R-sequence of length 2 contained in I. Let $c/d \in T(R)$ such that d is a regular element in R. If $(c/d)I \subseteq R$, then $ca = dr_1$ and $cb = dr_2$ for some $r_1, r_2 \in R$. Thus $br_1 = ar_2$ and hence $r_1 = ae$ for some $e \in R$. From the equation $ca = dae$ we have $c/d = e \in R$.

Define *depth* $R \leq 1$ to mean that the ring R does not contain an R-sequence of length ≥ 2. Rings of depth ≤ 1 are common in multiplicative ideal theory. For example all Prüfer domains have this property.

Theorem 19.4 If R is a ring such that $S = (\mathcal{U}_2)$, then depth $R \leq 1$. If R is Noetherian, then the converse holds.

PROOF: Assume that $\{a,b\}$ is an R-sequence of length two in R. Since $R[X] \setminus \cup M_\beta[X] = S = (\mathcal{U}_2) \subseteq R[X] \setminus \cup_{P \in \mathcal{P}_2} P[X]$, there is some $P \in \mathcal{P}_2$ such that $a + bX \in P[X]$. Hence $I = (a,b) \subseteq P$ which implies that $I^{-1} \neq R$, Corollary 19.2. This contradicts Lemma 19.3.

Assume that R is a Noetherian ring of depth ≤ 1. Let M be a regular maximal ideal of R (if one exists) and let a be a regular element in M. We claim that M is an associated prime ideal of (a). If not, $M \not\subseteq \cup_{i=1}^n P_i$ where $P_1, \ldots, P_n$ is the set of maximal associated prime ideals of (a). Choose $b \in M \setminus \cup_{i=1}^n P_i$ and note that $\{a,b\}$ forms an R-sequence in M. This contradiction substantiates our claim. Hence $M = (a : c)$ for some $c \in R$ [ZSI, p. 210], and therefore $M \in \mathcal{P}_2$. If $f \in (\mathcal{U}_2)$, then $f \notin P[X]$ for each $P \in \mathcal{P}_2$. In particular $f \notin M[X]$ for each regular maximal ideal M of R. On the other hand, since f is a regular element in $R[X]$, $c(f)$ is a regular ideal in R and therefore cannot be contained in any maximal ideal of R consisting entirely of zero divisors. Thus $f \in S$.

20. Kronecker Function Rings

Each ring in this section is a Marot ring with Property A. We construct Kronecker function rings with respect to $*$-operations. These constructions parallel the constructions for integral domains as given in [G, Section 32]; hence we only sketch the details.

Recall that $\mathcal{F}(R)$ is the set of R-submodules of $T(R)$. A *$*$-operation* on $\mathcal{F}(R)$ is a mapping $A \to A^*$ of $\mathcal{F}(R)$ into $\mathcal{F}(R)$ such that for each $a \in T(R)$ and for each $A, B, C \in \mathcal{F}(R)$:

(1) $aA^* \subseteq (aA)^*$; and if a is regular, $aA^* = (aA)^*$.
(2) $A \subseteq A^*$ and $A \subseteq B$ imply $A^* \subseteq B^*$.
(3) $(A^*)^* = A^*$.

The $*$-operation is (e.a.b.) if it also satisfies

(4) If A, B, and C are finitely generated and if A is regular then, $(AB)^* \subseteq (AC)^*$ implies $B^* \subseteq C^*$.

Lemma 20.1 Let $A \to A^*$ be a $*$-operation on $\mathcal{F}(R)$. Then for all $A, B \in \mathcal{F}(R)$:

(1) $(A+B)^* = (A^* + B^*)^*$.
(2) $(AB)^* = (AB^*)^* = (A^*B^*)^*$.

PROOF: See [G, pp. 393–394].

Call A a $*$-*ideal* if $A = A^*$ and say that A is $*$-*finite* in case there exists a finitely generated fractional ideal B such that $A^* = B^*$.

Lemma 20.2 If $*$ is an (e.a.b.) operation on $\mathcal{F}(R)$ and if $f, g \in R[X]$ such that g is regular, then $c(fg)^* = [c(f)c(g)]^*$.

PROOF: See [G, p. 398].

Theorem 20.3 Let R be a ring and $*$ be an (e.a.b.) operation on $\mathcal{F}(R)$. Define

$$R^* = \{f/g : f, g \in R[X],\ g \text{ regular, and } c(f)^* \subseteq c(g)^*\}$$

Then:

(1) R^* is a Bezout ring.
(2) $T(R^*) = T(R[X])$ and $R^* \cap T(R) = R$.
(3) If A is a finitely generated regular ideal of R, then $AR^* \cap T(R) = A^*$.

PROOF: The Marot property and Property A may be used in conjunction with Lemmas 20.1 and 20.2 to modify [G, pp. 398–400].

Several comments are in order. (1) Property A is needed to ensure that the ring R^* is well defined. (2) The Marot property is used to prove that R^* is a Bezout ring. (If A is a finitely generated regular ideal of R^*, then A may be generated by finitely many regular elements, Theorem 7.1. Hence, to show that R^* is a Bezout ring it suffices to show that if a, b are regular elements in R^*, then (a, b) is a principal ideal. The proof of this is exactly as in [G, p. 399].) (3) The ideal A, in the last part of Theorem 20.3, could have been assumed to be dense. But in rings with Property A, finitely generated dense = finitely generated regular.

Corollary 20.4 If a ring R admits an (e.a.b.) operation, then R is integrally closed.

The ring R^*, as defined in Theorem 20.3, is called the *Kronecker function ring of R with respect to* $*$.

Before continuing the discussion of $*$-operations, a theorem on extending valuations from rings to polynomial rings must be established. This is done in Theorem 20.7. Two lemmas are needed. Notice that the proofs of the next three results are of a much different character than the corresponding proofs for integral domains. Some of the valuations in the following lemmas are defined on rings other than total quotient rings.

Lemma 20.5 Let V be a valuation ring with corresponding valuation v. Define v' on $T(V)[X]$ by $v'(\sum_{i=0}^{n} a_i X^i) = \min\{v(a_i) : a_i \neq 0\}$. Then v' is a valuation on $T(V)[X]$ which extends v.

PROOF: We show that v' satisfies parts (a), (b), and (c) of the definition of valuation; see Section 5.

Let $f, g \in T(V)[X]$. We claim that $v'(fg) = v'(f) + v'(g)$. First assume that $fg = 0$. It suffices to prove that either $v'(f) = \infty$ of $v'(g) = \infty$. If one of the factors is zero, there is nothing to prove. Hence we suppose that f and g are both nonzero. By definition of v',

$$v'(f) + v'(g) = \min\{v(a_i)\} + \min\{v(b_j)\}$$

where $f = a_0 + a_1X + \cdots + a_mX^m$ and $g = b_0 + b_1X + \cdots + b_nX^n$. If $v(a_i) = \infty$ for all i, then $v'(f) = \infty$, so suppose that there is at least one coefficient of f with value $< \infty$. Proceed by induction on the degree of g.

If $\deg g = 0$, then $g = c \in T(V)$. Hence $ca_i = 0$ for all i. Let k be the first integer such that $v(a_k) < \infty$. Then $\infty = v(ca_k) = v(c) + v(a_k)$ which implies that $v(c) = v'(g) = \infty$. For the induction hypothesis assume that f has at least one coefficient whose value under v is less than ∞, $fg = 0$, and if $\deg g < n$ then each coefficient of g has infinite value under v.

If $b_0 = 0$, then the degree of g may be reduced by at least one and the result follows by the induction hypothesis. So in addition we may assume that $b_0 \neq 0$. Since $a_0b_0 = 0$, $a_0gf = a_0X(b_1 + \cdots + b_nX^{n-1})f = 0$. There are two cases to consider, when $a_0g = 0$ and when $a_0g \neq 0$. The two cases may be reduced to the same point, from which this part of the proof can then be completed. If $a_0g = 0$, then $a_0b_j = 0$ for each j. Thus $v(a_0) = \infty$ or $v(b_j) = \infty$ for $j = 0, \ldots, n$. The second case is when $a_0g \neq 0$. Thus $a_0(b_1 + \cdots + b_nX^{n-1}) \neq 0$. The induction hypothesis yields $v(a_0b_j) = \infty$ for $j = 1, \ldots, n$. Once again $v(a_0) = \infty$ or $v(b_j) = \infty$ for $j = 0, 1, \ldots, n$. If $v(a_0) < \infty$, then $v'(g) = \infty$. Assume that $v(a_0) = \infty$ and that there exists a coefficient of g with v-value $< \infty$. Choose k and p to be the smallest integers such that $v(a_k) < \infty$ and $v(b_p) < \infty$, respectively. Consider $c_{k+p} = \sum_{i+j=k+p} a_ib_j$, the coefficient of X^{k+p} in fg. Because $fg = 0$, $v(c_{k+p}) = \infty$. But $v(c_{k+p}) = v(a_k) + v(b_p) < \infty$, a contradiction. Therefore

$v(b_j) = \infty$ for all j and hence $v'(g) = \infty$. To summarize: If $fg \neq 0$, then $v'(f) + v'(g) = v'(fg)$. To complete the proof of part (a) (of the definition of a valuation) for the function v', we must show that if $fg \neq 0$, then $v'(fg) = v'(f) + v'(g)$. Consider $fg = 0$. Then $v'(fg) = \min\{v(c_k) : fg = \sum_{k=0}^{m+n} c_k X^k$ and $c_k = \sum_{i+j=k} a_i b_j\}$. Clearly $v(a_i b_j) = v(a_i) + v(b_j) \geq v'(f) + v'(g)$, for each i and j. Therefore $v'(fg) \geq v'(f) + v'(g)$. Let k and p be the smallest integers for which $v(a_k) = v'(f)$ and $v(b_p) = v'(g)$. The coefficients of X^{k+p} in fg is $a_k b_p + \sum a_w b_s$ where $w + s = k + p, w \neq k$, and $s \neq p$. We have $v(a_w b_s) = v(a_w) + v(b_s) > v(a_k) + v(b_p) = v'(f) + v'(g)$. Therefore $v'(fg) \leq v'(f) + v'(g)$. This completes the proof of part (a) of the definition. It is clear that parts (b) and (c) hold for v'. Finally the mapping v' of $T(V)[X]$ to the value group of v is surjective. Therefore v' is a valuation.

Lemma 20.6 Let v be a valuation on the ring R and let (R_v, P_v) be a valuation pair of R with respect to v. Assume that for each regular element $r \in R$, $v(r) < \infty$. Then there exists a unique extension v^* of v to $T(R)$ defined by $v^*(x/y) = v(x) - v(y)$. In addition, if (V, M) is the valuation pair of v^*, then $V \cap R = R_v$ and $M \cap R = P_v$.

PROOF: The hypothesis that each regular element of R has finite v-value is needed to show that v^* is well defined. The rest of the proof is standard.

Theorem 20.7 Let v be a valuation on a total quotient ring and let V be the corresponding valuation ring for v. Then v extends to a valuation on $T(V[X])$ if and only if for any finite subset E of $v^{-1}(\infty)$, $\operatorname{Ann} E \neq (0)$.

PROOF: ($\Rightarrow$): Let v^* be the extension of v to $T(V[X])$, let $\{a_0, \ldots, a_n\} \subseteq v^{-1}(\infty)$ and define $f = a_0 + a_1 X + \cdots + a_n X^n$.

Then

$$\begin{aligned} v^*(f) &\geq \min\{v^*(a_iX^i)\} \\ &= \min\{v(a_i) + iv^*(X)\} = \infty \end{aligned}$$

If $\operatorname{Ann}\{a_i\}_{i=0}^n = (0)$, then f is regular, which implies that $v^*(f) < \infty$. Therefore $\operatorname{Ann}\{a_i\}_{i=0}^n \neq (0)$.

($\Leftarrow$): By Lemma 20.5, v may be extended to a valuation v' on $T(V)[X]$ by the formula $v'(b_0 + b_1X + \cdots + b_mX^m) = \min\{v(b_i)\}$. If $f = a_0 + a_1X + \cdots + a_nX^n \in T(V)[X]$ such that $v'(f) = \infty$, then $\{a_0, \ldots, a_n\} \subseteq v^{-1}(\infty)$. From the hypothesis we conclude that f is a zero divisor. Hence if f is regular, then $v'(f) < \infty$. By Lemma 20.6, v' may be extended to a valuation v^* on $T(V[X])$.

Corollary 20.8 If V is a valuation subring of $T(V)$ with corresponding valuation v and if V has Property A, then v may be extended to a valuation on $T(V[X])$.

PROOF: If x is a regular element in $v^{-1}(\infty)$, then $0 = v(x) + v(x^{-1}) = \infty - \infty$, a contradiction. Hence $v^{-1}(\infty) \subseteq Z(V)$. If E is a finite subset of $v^{-1}(\infty)$, then the ideal generated by $E \subseteq v^{-1}(\infty) \subseteq Z(V)$. Property A implies that $\operatorname{Ann} E \neq (0)$. By Theorem 20.7, v may be extended to a valuation on $T(V[X])$.

The valuation v^*, constructed above, is called the *natural extension* of v to $T(V[X])$.

We return to the investigations of $*$-operations. If R is an integrally closed ring, then by the Marot property $R = \cap V_\alpha$ where $\{V_\alpha\}$ is the set of valuation overrings of R (Theorem 9.3). For A in $\mathcal{F}(R)$ define the *completion* of A by $A' = \cap AV_\alpha$. Say that A is *complete* if $A = A'$.

Theorem 20.9 Let R be an integrally closed ring. For $A \in \mathcal{F}(R)$ let $(A^{-1})^{-1} = A_v$. Then:

(1) The mapping $A \to A'$ is an (e.a.b.) $*$-operation on $\mathcal{F}(R)$.

(2) The mapping $A \to A_v$ is a $*$-operation on $\mathcal{F}(R)$.

PROOF: Both of these proofs are easy generalizations of the corresponding proofs for integral domains that may be found in [G].

The $*$-operation defined by $A \to A'$ is called the *b-operation* on R. Following [G] we let R^K denote the Kronecker function ring of R with respect to this operation. The mapping $A \to A_v$ is called the *v-operation.* We are especially interested when this operation is an (e.a.b.) $*$-operation. In this case we say that the ring R is a *v-ring* and denote the corresponding Kronecker function ring by R^v. The rings R^K and R^v are the subjects of the next two sections.

21. Kronecker Function Rings and $R(X)$

Each ring in this section is a Marot ring with Property A. We characterize Prüfer rings in terms of R^K and $R(X)$.

The lemma below generalizes a well-known result for integral domains.

Lemma 21.1 Let R be an integrally closed ring, let P be a prime ideal of R, and let t be regular element in $T(R)$ that satisfies a polynomial $f \in R[X] \setminus P[X]$. Then t or t^{-1} is in $R_{[P]}$.

PROOF: Let $f = a_0 + a_1X + \cdots + a_nX^n$. If $a_0 \notin P$, then $a_0(t^{-1})^n + a_1(t^{-1})^{n-1} + \cdots + a_n = 0$, and so a_0t^{-1} is integral over R, hence in R. Thus $t^{-1} \in R_{[P]}$. Similarly if $a_n \notin P$, then $a_nt \in R$ and $t \in R_{[P]}$.

We assume that $a_0, a_n \in P$ and choose i such that $a_n, \ldots, a_{i+1} \in P$, $a_i \notin P$. Then $a_nt^n + \cdots + a_it^i = -a_{i-1}t^{i-1} - \cdots - a_0$. Multiplying this equation by $(t^{-1})^i$ yields

$$b = a_nt^{n-i} + \cdots + a_i = -a_{i-1}t^{-1} - \cdots - a_0(t^{-1})^i$$

Since $b \in R[t] \cap R[t^{-1}]$, b is integral over R [G, p. 227] and therefore $b \in R$. If $b \in P$, then $a_i - b \notin P$ and $a_n t^{n-i} + \cdots + (a_i - b) = 0$. By the first case, $t^{-1} \in R_{[P]}$. If $b \notin P$, then $a_0(t^{-1})^i + \cdots + a_{i+1}t^{-1} + b = 0$ which, also by an earlier case, implies that $t \in R_{[P]}$.

Theorem 21.2 Let R be an integrally closed ring and let N be a multiplicatively closed subset of $R[X]$. If each regular prime ideal of $R[X]_{(N)}$ is extended from a prime ideal of R, then $R[X]_{(N)}$ is a Bezout ring.

PROOF: Our first goal is to show that $R[X]_{(N)}$ is a Prüfer ring. Let $\mathcal{M}$ be a maximal ideal of $R[X]_{(N)}$. We must prove that $[R[X]_{(N)}]_{[\mathcal{M}]} (= [R[X]_{(N)}]_{(\mathcal{M})})$ is a valuation ring. If $\mathcal{M}$ consists entirely of zero divisors, then $[R[X]_{(N)}]_{(\mathcal{M})} = T(R[X])$ is a valuation ring. So we may assume that $\mathcal{M}$ is a regular maximal ideal of $R[X]_{(N)}$. Let P be a prime ideal of R that extends to $\mathcal{M}$. Then $\mathcal{M} = P[X]R[X]_{(N)}$ and so $P[X] \cap N = \emptyset$. Thus

$$R[X]_{(P[X])} = [R[X]_{(N)}]_{(PR[X]_{(N)})} = [R[X]_{(N)}]_{(\mathcal{M})}$$

Since $R[X]_{(R \setminus P)} = R_{(P)}[X]$,

$$\begin{aligned} R[X]_{(P[X])} &= (R[X]_{(R\setminus P)})_{[P[X]R[X]_{(R\setminus P)}]} \\ &= [R_{(P)}[X]]_{(PR_{(P)}[X])} \end{aligned} \tag{21a}$$

Claim 1. $R_{(P)}$ is a valuation ring. Let $t = a/b$ be a regular element of $T(R)$. Define $\varphi : R[X] \to R[t]$ to be the homomorphism that maps X to t and acts as the identity on R. Let $\mathcal{Q}$ be a minimal prime ideal of $\mathcal{B} = \ker \varphi$. The ideal $\mathcal{B}$, and hence $\mathcal{Q}$, is a regular ideal since it contains the polynomial $bX - a$. It is easily seen that $\mathcal{Q} \cap R \subseteq \varphi(\mathcal{Q}) \cap R \subseteq Z(R)$.

Recall that (N) is the set of regular elements in N. Suppose that $\mathcal{Q} \cap (N) = \emptyset$. By the hypothesis there is a prime ideal M of R such that $\mathcal{Q}R[X]_{(N)} = M[X]R[X]_{(N)}$. Therefore $\mathcal{Q} = M[X]$.

Then $\mathcal{Q} \cap R = M$ must be a regular prime ideal of R. To see this last statement note that $aX - b \in M[X]R[X]_{(N)} = MR[X]_{(N)}$, so $aX - b = m_1 f_1/g + \cdots + m_t f_t/g$ where $m_i \in M$, $f_i \in R[X]$, and $g \in (N)$. Then $(aX - b)g = m_1 f_1 + \cdots + m_t f_t$ is a regular element of $R[X]$. By Property A, the dense ideal $(m_1, \ldots, m_t)$ is a regular ideal in $M = \mathcal{Q} \cap R$. This contradiction yields $\mathcal{Q} \cap (N) \neq \emptyset$.

Return to the prime ideal $P = R \cap \mathcal{M}$ that was described in the first paragraph of this proof. By the preceding paragraph, $\mathcal{B} \not\subseteq P[X]$. Choose $f(X) \in \mathcal{B} \setminus P[X]$. Since $f(t) = 0$, Lemma 21.1 implies that t or t^{-1} is in $R_{[P]}$. By Theorem 7.6 $R_{[P]} = R_{(P)}$, and by Theorem 7.7 $R_{(P)}$ is a valuation ring.

Claim 2. $R[X]_{(P[X])}$ is a valuation ring. (This will establish that $R[X]_{(N)}$ is a Prüfer ring.) If $P \subseteq Z(R)$, then by Property A $P[X] \subseteq Z(R[X])$, and hence $R[X]_{(P[X])} = T(R[X])$ is a valuation ring. Therefore it is sufficient to assume that P is a regular prime ideal of R. Let v be the valuation corresponding to $R_{(P)}$. Since R has Property A, v may be extended to a valuation v^* on $T(R[X])$—the natural extension of v to $T(R[X])$, Corollary 20.8. For an element $a_0 + a_1X + \cdots + a_nX^n \in R_{(P)}[X]$, $v^*(a_0 + a_1X + \cdots + a_nX^n) = \min\{v(a_i)\}$. We show that the valuation ring of v^* is $R[X]_{(P[X])}$. Let $f/g \in T(R[X])$ where $f, g \in R[X]$ and g is regular. Write $f = a_0 + a_1X + \cdots + a_nX^n$ and $g = b_0 + b_1X + \cdots + b_mX^m$ and assume that $v^*(f/g) \geq 0$. There is a smallest integer k such that $v(b_k) \leq \min_{i,j}\{v(a_i), v(b_j)\}$. Each overring of a Marot ring is a Marot ring (Corollary 7.3); hence $R_{(P)}$ is a Prüfer valuation ring (Theorem 7.7). Theorem 7.9 implies the existence of a regular element $c \in R_{(P)}$, such that $v(c) = v(b_k)$. Rewrite f/g as

$$\frac{f}{g} = \frac{(a_0/c) + (a_1/c)X + \cdots + (a_n/c)X^n}{(b_0/c) + \cdots + (b_k/c)X^k + \cdots + (b_m/c)X^m} \tag{21b}$$

Note that each $v(a_i/c)$ and $v(b_j/c)$ are nonnegative, so the numerator and denominator of (21b) are in $R_{(P)}[X]$. Since P is a regular prime ideal of R, $PR_{(P)} = \{x \in T(R) : v(x) > 0\}$, Theorem 6.5. Now $v(b_k/c) = 0$ implies that $b_k/c \notin PR_{(P)}$. It follows that $f/g \in R[X]_{(P[X])}$, (21a). The elements of $R[X]_{(P[X])}$ have nonnegative v^*-value. Therefore $R[X]_{(P[X])}$ is the valuation ring of v^*.

Claim 3. $R[X]_{(N)}$ is a Bezout ring. Let $\mathcal{I}$ be a finitely generated regular ideal of $R[X]_{(N)}$. By Theorem 7.1, $\mathcal{I}$ may be generated by finitely many regular elements and furthermore these elements may be picked in $R[X]$. Hence we need only prove that if f, g are regular elements in $R[X]$, then $(f,g)R[X]_{(N)}$ is a principal ideal. Choose $h \in (f,g)R[X]$ such that $c(h) = c(f) + c(g)$. We prove that, for each maximal ideal $\mathcal{M}$ of $R[X]_{(N)}$, $(f,g)(R[X]_{(N)})_{\mathcal{M}} = h(R[X]_{(N)})_{\mathcal{M}}$; hence $(f,g)R[X]_{(N)} = hR[X])_{(N)}$. Let $\mathcal{M}$ be an arbitrary maximal ideal of $R[X]_{(N)}$. Assume that $\mathcal{M}$ is regular; for if not, we certainly have $(f,g)(R[X]_{(N)})_{\mathcal{M}} = h(R[X]_{(N)})_{\mathcal{M}}$. Let P be the prime ideal of R that extends to $\mathcal{M}$. By (21a) and Claim 1, we may assume that (R,P) is a Marot valuation pair and that $[R[X]_{(N)}]_{\mathcal{M}} = R[X]_{(P[X])}$. We prove that $(f,g)R[X]_{(P[X])} = hR[X]_{(P[X])}$. Suppose that v is the valuation associated with the valuation pair (R,P). If $f = b_0 + b_1X + \cdots + b_nX^n$, then $c(f) = (b_0, \ldots, b_n)$ is a regular ideal and $c(f)$ is generated by finitely many regular elements $\{a_1, \ldots, a_t\}$. Out of this particular set of generators choose the first regular element a of minimum v-value. It is easy to see that $c(f) = (a)$. By Lemma 14.6, there exists some $f_1 \in R[X]$ such that $f = af_1$ and $c(f_1) = R$. Similarly, there exists some $g \in R[X]$ and regular $b \in R$ such that $g = bg_1$ and $c(g_1) = R$. If $(c) = (a,b)$, then $h = ch_1$ where $c(h_1) = R$. Consequently

$$\begin{aligned}(f,g)R[X]_{(P[X])} &= (a,b)R[X]_{(P[X])} = cR[X]_{(P[X])} \\ &= hR[X]_{(P[X])}\end{aligned}$$

Theorem 21.3 A necessary and sufficient condition for a ring R to be Prüfer is that every ideal of R is complete.

PROOF: Assume that R is a Prüfer ring. The Marot property implies that R is the intersection of the set of valuation rings $\{V_\alpha\}$ between R and $T(R)$, Theorem 9.3. For each maximal ideal M_β of $R, R_{[M_\beta]}$ is a valuation ring. By Theorem 6.1, $I \subseteq \cap IV_\alpha \subseteq \cap IR_{[M_\beta]} = I$, thus $I = \cap IV_\alpha$.

We establish the converse by showing that condition (4) of Theorem 6.2 holds for R. Let $IJ = IK$ where I, J, and K are ideals of R, with I finitely generated and regular. Then $(IV)(JV) = (IV)(KV)$ for each valuation overring V of R. Since V is a Marot ring, IV is a regular principal ideal and is therefore invertible. Hence $JV = KV$ for all V. But each ideal of R is complete; thus $J = K$.

Note that Property A is not needed in the preceding theorem. We are ready for the main result of this section.

Theorem 21.4 Let R be an integrally closed ring. Then the following conditions are equivalent:

(1) R is a Prüfer ring.
(2) $R(X) = R^K$.
(3) $R(X)$ is a Bezout ring.
(4) R^K is a regular quotient ring of $R[X]$.
(5) Each regular prime ideal of $R(X)$ is a contracted prime ideal of R^K.
(6) Each regular prime ideal of $R(X)$ is an extension of a prime ideal of R.

PROOF: (1) $\Rightarrow$ (2): That $R(X) \subseteq R^K$ follows directly from the definitions of the two rings involved.

Let $z = f/g \in R^K$ where $f, g \in R[X]$, g is regular, and $c(f)' \subseteq c(g)'$. The ideal $c(g)$ is a finitely generated regular ideal of the Prüfer ring R, and is therefore invertible. Say $c(g)^{-1} = (u_0, \ldots, u_n), u_i \in T(R)$. The polynomial $u = u_0 + u_1X + \cdots + u_nX^n$ is a regular element in $T(R)[X]$. For each

i write $u_i = t_i/r$ where $t_i, r \in R$, and r is regular. If $h = t_0 + t_1X + \cdots + t_nX^n$, then $u = r^{-1}h$. Clearly h is a regular element of $R[X]$. From Lemma 20.2 and Theorem 21.3, $c(gu) = r^{-1}c(gh) = c(g)r^{-1}c(h) = c(g)c(g)^{-1} = R$. Therefore $z = fu/gu \in R(X)$.

(2) $\Rightarrow$ (3): Theorem 20.3.

(3) $\Rightarrow$ (1): Theorem 18.10.

(2) $\Rightarrow$ (4) and (2) $\Rightarrow$ (5): Clear.

(4) $\Rightarrow$ (2): Apply the proof of (4) $\Rightarrow$ (2) of Theorem 33.4 [G, p. 414].

(5) $\Rightarrow$ (6): Assume that $\mathcal{P}$ is a regular prime ideal in R^K and that $P = \mathcal{P} \cap R$. Then $P[X] \subseteq \mathcal{P} \cap R[X]$. If f is a regular element in $\mathcal{P} \cap R[X]$, write $f = a_0 + a_1X + \cdots + a_nX^n$. Note that $fR^K \subseteq c(f)R^K$. On the other hand, for each i, $a_i/f \in R^K$, since $c(a_i) \subseteq c(f)$. Thus $fR^K = c(f)R^K$. We have proved that $f \in P[X]$ and therefore $P[X] = \mathcal{P} \cap R[X]$.

Let $\mathcal{Q}$ be a regular prime ideal of $R(X)$. We are looking for a prime ideal P in R such that $PR(X) = \mathcal{Q}$. If $\mathcal{P}$ is the prime ideal of R^K such that $\mathcal{P} \cap R(X) = \mathcal{Q}$, let $P = \mathcal{P} \cap R$. Then

$$\mathcal{Q} \cap R[X] = \mathcal{P} \cap R[X] = P[X]$$

Since $R(X)$ is a quotient ring of $R[X]$, $\mathcal{Q} = PR(X)$.

(6) $\Rightarrow$ (3): Theorem 21.2.

Theorem 21.4 gives a sharper version of Theorem 18.10, when R is a Marot ring with Property A.

22. Kronecker Function Rings and $R[X]_{(\mathcal{U}_2)}$

Each ring in this section is a Marot ring with Property A. We characterize Prüfer v-multiplication rings in terms of the Kronecker function ring R^v. From this a considerable amount of information may be obtained concerning Krull domains, GCDs, etc. Recall that a v-ring is a ring in which the v-operation is (e.a.b.). It is for this class of rings that the Kronecker function ring R^v may be defined.

We need the definitions and results concerning $\mathcal{F}^*(R)$, div A, and $\mathcal{D}(R)$ from Section 8. An element $A \in \mathcal{F}(R)$ is of *finite v-type* if there exists a finitely generated $B \in \mathcal{F}^*(R)$ such that $A_v = B_v$. The divisor class div A is of *finite type* if A is of finite v-type. Let H be the subset of $\mathcal{D}(R)$ consisting of divisor classes of finite type; then H may or may not be a group. There is an example of a completely integrally closed domain D (that is, $\mathcal{D}(D)$ is a group) such that H is not a group [G, p. 429]. If H is a group, then R is called a *Prüfer v-multiplication ring* (PVMR). Note that we are not assuming that $\mathcal{D}(R)$ is necessarily a group in order that R be a PVMR.

The proofs of the next two results are the same as the corresponding proofs for domains; see [G, Section 34].

Theorem 22.1 A ring R is a v-ring if and only if each divisor class of finite type has an inverse in $\mathcal{D}(D)$.

Theorem 22.2 A PVMR is a v-ring.

The *group of divisibility* G of a ring R with zero divisors is defined as follows: If $T(R)^*$ is the group of units in $T(R)$ and if $\mathcal{U}$ is the subgroup of $T(R)^*$ consisting of all units in R, then $G = T(R)^*/\mathcal{U}$. We order G by defining $x\mathcal{U} \leq y\mathcal{U}$ if and only if $yx^{-1} \in R$. Since R^v is a Bezout ring, G is order isomorphic to S, S being the group of regular principal fractional ideals of R, via the mapping $(z) \to z\mathcal{U}$. Hence,

Theorem 22.3 The group H is order isomorphic to the group of regular principal fractional ideals of R^v via the map div $A \to AR^v$, where $A \in \mathcal{F}^*(R)$ is finitely generated and regular.

To reiterate, the proofs of the above results may be found in [G]. (Actually in [G] the proofs are for integral domains, but with simple modifications these proofs carry over to rings with zero divisors.)

Lemma 22.4 Assume that R is an additively regular ring. If $\mathcal{Q}$ is a prime ideal of the polynomial ring $R[X]$ such that $R[X]_{(\mathcal{Q})}$ is a valuation ring and $(\mathcal{Q} \cap R)R[X] \subset \mathcal{Q}$, then $\mathcal{Q} \cap R \subseteq Z(R)$.

PROOF: Let $P = \mathcal{Q} \cap R$; then $R_P \subseteq R[X]_{(\mathcal{Q})} \cap T(R)$. On the other hand let $f/g \in R[X]_{(\mathcal{Q})} \cap T(R)$ where $f, g \in R[X]$ and g is a regular element not in $\mathcal{Q}$. Then $g = r_0 + \cdots + r_i X^i + \cdots + r_n X^n$, where $r_i \notin \mathcal{Q}$. Since $f/g = z \in T(R)$, $zg = f \in R[X]$ and $zr_i \in R$. Hence $z \in R_{[P]} = R_{(P)}$. This proves that $R_{(P)} = R[X]_{(\mathcal{Q})} \cap T(R)$ is a valuation ring.

If $N = R \setminus P$, then $R[X]_{(\mathcal{Q})} = (R[X]_{(N)})_{(\mathcal{Q}_{(N)})} = R_{(P)}[X]_{(\mathcal{Q}R_{(P)}[X])}$. Define R_0 to be the valuation ring $R_{(P)}$ and $\mathcal{Q}_0$ to be $\mathcal{Q}R_{(P)}[X]$. Then $\mathcal{Q}_0 \cap R = \mathcal{Q} \cap R$ and $(\mathcal{Q}_0 \cap R)R_0[X] \subset \mathcal{Q}_0$.

We will prove that $\mathcal{Q}_0 \cap R \subseteq Z(R)$. By using the notation in the paragraph preceding Theorem 18.11, $\mathcal{Q}_0 = \langle \mathcal{Q}_0 \cap R_0, a(X) \rangle = \{g(X) \in R_0[X] : a(X) \text{ divides } \bar{g}(X)\}$ where $\bar{g}(X)$ is the result of reducing the coefficients of $g(X)$ modulo $\mathcal{Q}_0 \cap R_0$ and $a(X)$ is a monic irreducible polynomial in $F[X]$ (F = quotient field of $R_0/(\mathcal{Q}_0 \cap R_0)$). Write

$$a(X) = \bar{a}_0/\bar{b} + (\bar{a}_1/\bar{b})X + \cdots + (\bar{a}_{n-1}/\bar{b})X^{n-1} + X^n$$

Let $f(X) = a_0 + a_1 X + \cdots + a_{n-1}X^{n-1} + bX^n \in R_0[X]$ and note that $\bar{f}(X) = \bar{b}a(X)$, so that $f(X) \in \mathcal{Q}_0$. Clearly $b \notin \mathcal{Q}_0 \cap R_0$.

Assume that $\mathcal{Q}_0 \cap R_0$ contains a regular element r. Since R, and hence R_0, is additively regular, there exists $u \in R_0$ such that $b + ur$ is regular. Define $g(x) = a_0 + a_1 X + \cdots + a_{n-1}X^{n-1} + (b + ur)X^n$. Then $\bar{b} = \overline{b + ur}$ so $\bar{f}(X) = \bar{g}(X)$, which implies that $g(X) \in \mathcal{Q}_0 \setminus (\mathcal{Q}_0 \cap R_0)R_0[X]$. Furthermore the monic polynomial $h(X) = g(X)/(b + ur) \in \mathcal{Q}_0 \setminus (\mathcal{Q}_0 \cap R)R_0[X]$, since $R_0 = R_{(\mathcal{Q} \cap R)}$. For any regular $z \in \mathcal{Q}_0 \cap R$ we have $z \in h(X)R_0[X]_{(\mathcal{Q}_0)}$, since $R_0[X]_{(\mathcal{Q}_0)}$ is a valuation ring, $h(X)$ is regular, and $h(X) \notin zR_0[X]_{(\mathcal{Q}_0)}$. Write $z = h(X)f_1(X)/f_2(X)$ where $f_1(X), f_2(X) \in R_0[X]$ and $f_2(X)$ is a regular element not in $\mathcal{Q}_0$. Thus $f_2(X) = z^{-1}h(X)f_1(X)$. By [G, p. 93], $z^{-1}f_1(X) \in R_0[X]$. We conclude that $f_2(X) \in h(X)R_0[X] \subseteq \mathcal{Q}_0$, a contradiction. This shows that $\mathcal{Q}_0 \cap R_0 \subseteq Z(R_0)$. It easily follows that $\mathcal{Q} \cap R \subseteq Z(R)$.

We characterize the additively regular v-rings which are PVMRs. Recall that the additively regular hypothesis is slightly weaker than the Marot hypothesis.

Theorem 22.5 Let R be an additively regular v-ring. Then the following conditions are equivalent:

(1) R is a PVMR.
(2) $R[X]_{(\mathcal{U}_2)}$ is a Bezout ring.
(3) Each regular prime ideal if $R[X]_{(\mathcal{U}_2)}$ is extended from a prime ideal of R.
(4) $R^v = R[X]_{(\mathcal{U}_2)}$.
(5) R^v is a flat $R[X]$-module.
(6) Each valuation overring of R^v is of the form $R[X]_{(P[X])}$, where $R_{(P)}$ is a valuation ring.
(7) $R[X]_{(\mathcal{U}_2)}$ is a Prüfer ring.

PROOF: (1) $\Leftrightarrow$ (4): Assume that the ring R is a PVMR. Clearly $R[X]_{(\mathcal{U}_2)} \subseteq R^v$. Let $z = f/g$ where $f, g \in R[X]$, g is regular and $c(f)_v \subseteq c(g)_v$. There exists a finitely generated regular $A \in \mathcal{F}^*(R)$ such that $(c(g)A)_v = R$. Let $h_0/d, \ldots, h_n/d$ generate A with $h_i, d \in R$ and d regular. Define $h = h_0 + h_1X + \cdots + h_nX^n$ and note that $z = fh/gh = d^{-1}fh/ghd^{-1}$. We have $R = (c(g)A)_v = [c(ghd^{-1})]_v$ (Lemma 20.2). Thus $ghd^{-1} \in R[X]$. In fact, $ghd^{-1} \in (\mathcal{U}_2)$ (Theorem 19.1). Therefore $R[X]_{(\mathcal{U}_2)} = R^v$.

Conversely assume that $R^v = R[X]_{(\mathcal{U}_2)}$. Let A be a regular finitely generated fractional ideal of R and write $A = d^{-1}I$ where $I = (a_0, \ldots, a_n)$ is an integral ideal of R. Since R is a Marot ring, each a_i may be picked to be a regular element of I. Let $k = a_0 + a_1X + \cdots + a_nX^n \in R[X]$. Then $(a_0)_v = (a_0) \subseteq c(k)_v$, which implies that $a_0/k = a_0d^{-1}/kd^{-1} \in R^v$. So there exist $g, f \in R[X]$, g regular and $c(g)_v = R$ such that $a_0d^{-1}/kd^{-1} = f/g$. We have the chain of equalities

$$\begin{aligned}(a_0d^{-1})_v = (a_0d^{-1}) = (a_0d^{-1})c(g)_v = c(a_0gd^{-1})_v = c(fkd^{-1})_v \\ = [c(kd^{-1})_vc(f)_v]_v\end{aligned}$$

This implies that

$$R = (da_0^{-1})[c(kd^{-1})_vc(f)_v]_v = [c(kd^{-1})_vc(da_0^{-1}f)_v]_v$$

That is, A has an inverse in H. Therefore H is a group.

(4) ⇒ (6): Every proper valuation overring of R^v is of the form $R^v_{(\mathcal{Q})}$ where $\mathcal{Q}$ is a regular prime ideal of R^v, [LM, p. 248]. Therefore $R^v_{(\mathcal{Q})} = [R[X]_{(\mathcal{U}_2)}]_{(\mathcal{Q})} = R[X]_{(\mathcal{Q}\cap R[X])}$. Let $f = a_0 + a_1X + \cdots + a_nX^n$ be a regular element in $\mathcal{Q} \cap R[X]$. For each i, $a_i/f \in R^v$; hence $c(f)R^v \subseteq fR^v$. Since the other inclusion is obvious, $c(f)R^v = fR^v \subseteq \mathcal{Q}$. Hence $c(f) \subseteq \mathcal{Q} \cap R = P$ must be a regular prime ideal of R. By Lemma 22.4, $\mathcal{Q} \cap R[X] = P[X]$. Thus $R^v_{(\mathcal{Q})} = R[X]_{(P[X])}$ is a valuation ring. As we saw in the proof of Lemma 22.4, $R_{(P)} = R[X]_{(P[X])} \cap T(R)$ is a valuation ring.

(6) ⇒ (5): Let $\mathcal{M}$ be a regular maximal ideal in R^v. The hypothesis implies that $R^v_{(\mathcal{M})} = R[X]_{(P[X])}$ for some $P \in \operatorname{Spec} R$. Thus $\mathcal{M}R^v_{(\mathcal{M})} = P[X]R[X]_{(P[X])}$, which implies that $\mathcal{M}R^v_{(\mathcal{M})} \cap R[X] = P[X]$, whence $\mathcal{M} \cap R[X] = P[X]$. By [LM, p. 248], R^v is a flat $R[X]$-module.

(5) ⇒ (4): It is always true that $R^v \supseteq R[X]_{(\mathcal{U}_2)}$. Let $\mathcal{I}$ be an ideal in $R[X]$ such that $\mathcal{I}R^v = R^v$. Choose $f_1, \ldots, f_n \in \mathcal{I}$ such that $(f_1, \ldots, f_n)R^v = R^v$. If $m = 1 + \max\{\deg f_i\}$ and if

$f = f_1 + f_2X^m + \cdots + f_nX^{(n-1)m}$, then f is a regular element in $R[X]$ such that $fR^v = c(f)R^v$. Therefore f is a unit in R^v, forcing $c(f)^{-1} = R$. Thus $\mathcal{I} \cap (\mathcal{U}_2) \neq \emptyset$.

If $\mathcal{M}$ is a regular maximal ideal of $R[X]_{(\mathcal{U}_2)}$, the above argument implies that $\mathcal{M}R^v \subset R^v$. Hence each $[R[X]_{(\mathcal{U}_2)}]_{(\mathcal{M})} \supseteq R^v$ [LM, p. 248]. Therefore $R[X]_{(\mathcal{U}_2)} = \cap[R[X]_{(\mathcal{U}_2)}]_{(\mathcal{M})}$, Theorem 6.1.

(3) $\Rightarrow$ (2): Since R is an integrally closed ring, Theorem 21.2 is applicable.

(1) $\Rightarrow$ (3): Let $\mathcal{P}$ be a regular prime ideal of $R[X]_{(\mathcal{U}_2)} = R^v$ and let f be a regular element in $\mathcal{P}$. By Theorem 22.3, there exists a regular finitely generated fractional ideal A of R such that $AR^v = fR^v$. Then $A \subseteq A_v = AR^v \cap T(R) \subseteq R^v \cap T(R) = R$, hence A is a regular ideal of R contained in $\mathcal{P} \cap R$. Lemma 22.4 forces $(\mathcal{P} \cap R)R[X] = \mathcal{P} \cap R[X]$. Consequently $\mathcal{P} = (\mathcal{P} \cap R)R^v = (\mathcal{P} \cap R)R[X]_{(\mathcal{U}_2)}$.

(2) $\Rightarrow$ (7) and (7) $\Rightarrow$ (5) are clear.

We give some applications of Theorem 22.5 to the theory of integral domains.

Corollary 22.6 If D is either

(1) a GCD-domain,
(2) a Krull domain, or
(3) an integrally closed coherent ring,

then the equivalent conditions of Theorem 22.5 hold.

PROOF: It is well known that GCD-domains and Krull domains are Prüfer v-multiplication domains [G], hence they are PVMRs. It follows from [G, p. 432] that an integrally closed coherent ring is a PVMR.

When D is an integral domain the sets $\mathcal{U}_1$ and $\mathcal{U}_2$ coincide. In this case we write $\mathcal{U} = \mathcal{U}_1 = \mathcal{U}_2$. Similarly, denote $\mathcal{P}_1 = \mathcal{P}_2$ by $\mathcal{P}$. The preceding corollary says that if D is a Krull domain, then $D[X]_{\mathcal{U}}$ is a Bezout domain. More can be said.

Theorem 22.7 If D is a Krull domain, then $D[X]_{\mathcal{U}}$ is a principal ideal domain. If $D[X]_{\mathcal{U}}$ is a Krull domain, then D is a Krull domain.

PROOF: Let $\{P_\lambda\}$ be the set of height one prime ideals of D. We show that $\mathcal{P} = \{P_\lambda\} \cup \{(0)\}$. Certainly the right-hand side is contained in the left-hand side. Let M be a prime ideal of D of height ≥ 2. Choose a nonzero element $a \in M$ and let $P_1, \ldots, P_t$ be the complete set of height one prime ideals containing a. If $b \in M \setminus \cup P_i$, then a and b belong to no common height one prime ideal of D. By Theorem 8.17, $(a) = \cap Q_i$ where Q_i is P_i-primary. Since $b \notin \cup P_i$, and since $(a : b)b \subseteq (a) \subseteq Q_i$ for each i, we deduce that $(a : b) \subseteq Q_i$ for each i. Therefore $(a : b) = (a)$ which implies that $\{a, b\}$ is an R-sequence in M. Thus $I^{-1} = D$ where $I = (a, b)$, Lemma 19.3. By Corollary 19.2, $M \notin \mathcal{P}$.

Let $\mathcal{Q}$ be a nonzero prime ideal in $D[X]$ such that $\mathcal{Q} \cap \mathcal{U} = \varnothing$. If P is a nonzero member of $\mathcal{P}$, then P has height one and D_P is a discrete rank one valuation domain. By [G, p. 369], $P[X]$ is a height one prime ideal of $D[X]$. Thus if h is a nonzero element in $\mathcal{Q}$, then there are only finitely many $P_i \in \mathcal{P}$, say, $P_1, \ldots, P_t$, such that $h \in P_i[X]$ (for $D[X]$ is also a Krull domain). If $\mathcal{Q} \not\subseteq \cup P_i[X]$, choose $k \in \mathcal{Q} \setminus \cup P_i[X]$ and set $s = \deg h$. Then $h + kX^{s+1} \in \mathcal{Q}$. By definition of $\mathcal{U}$, there exists some $P' \in \mathcal{P}$ such that $h + kX^{s+1} \in P'[X]$. Hence $c(h) + c(k) \subseteq P'$, and therefore $h \in P'[X]$. There is some j such that $P' = P_j$ and $k \in P_j[X]$, a contradiction. This proves that $\mathcal{Q} \subseteq \cup P_i[X]$. Since $\mathcal{Q}$ is nonzero, $\mathcal{Q} = P_j[X]$ for an appropriate j between 1 and t. Thus $\mathcal{Q}_{\mathcal{U}}$ is a height one prime ideal in $D[X]_{\mathcal{U}}$. Therefore $D[X]_{\mathcal{U}}$ is a Bezout domain (Theorem 22.5) and $\dim D[X]_{\mathcal{U}} \leq 1$. This is enough to conclude that $D[X]_{\mathcal{U}}$ is a principal ideal domain, [G, p. 536].

The second statement follows from the easily proved fact that $D[X]_{\mathcal{U}} \cap T(D) = D$.

Notes

Section 14 The ring $R(X)$ and its elementary properties are from Nagata [N]. Quartararo and Butts studied μ-rings and proved (14.3), (14.4), and (14.5) in [95]. These rings were called u-rings in [95]. Since u-rings had been used in [G] in a different context, the above notation was adopted. A version of (14.7) appears in LeRiche's paper [65]. His proof was simplified and extended to $R(X)$ by D. D. Anderson, D. F. Anderson, and Markanda [6].

Section 15 The first four results are from D. D. Anderson [4]. That $\operatorname{Pic} R(X) = (0)$ (15.7), was proved independently by McDonald and Waterhouse [71], Ferrand [31], and by D. D. Anderson [5]. Anderson's proof is the one we use here.

Section 16 The results of this section are akin to those of Chapter III; and in particular to Section 13. Theorem 16.1 is from Gilmer and Hoffman [38]. Theorems 16.4 and 16.10 are due to Lucas [66]. The rest of this section comes from Akiba's two papers [1] and [2].

Section 17 The ring $R\langle X\rangle$ played an important role in Quillen's solution of Serre's conjecture [97]. This ring has been studied quite thoroughly. The dimension theory result (17.3) is from [65]. Gilmer and Heinzer prove (17.6)–(17.12) in [37]. Theorem 17.12 is certainly an interesting result for a book on commutative ring theory.

Section 18 [65] is the first paper to systematically study $R\langle X\rangle$. Results (18.3), (18.5), (18.9), and part of (18.12) are from this paper. Theorems (18.2), (18.6), (18.8), (18.10), and the completion of the work on (18.12) are in [6]. Theorem 18.13 is due to Dixon [26].

Section 19 Parts (1) and (2) of (19.1) are proved for integral domains by Tang [104]. Part 3 of (19.1), as well as the ring theory versions of (1) and (2) are by Huckaba and Papick [56]. Theorem 19.4 is also from [56]. For Noetherian rings, Theorem 19.4 gives a necessary and sufficient condition for $S = (\mathcal{U}_2)$. Actually, in the class of Noetherian rings, it is known when each pair of multiplicatively closed sets of (19a) are equal [56]. The corresponding characterizations for non-Noetherian rings are not known. Concerning (19a), there is an example in [56] that shows $\mathcal{U}_1 \subset (\mathcal{U}_2)$.

Section 20 The development of $*$-operations and Kronecker function rings for rings with zero divisors parallels the corresponding development for domains. Theorems (20.5), (20.7), and (20.8) are proved in [51].

Section 21 Arnold [13] gives the following characterizations of Prüfer domains in terms of overrings of polynomial rings.

Theorem ($*$) If D is an integrally closed domain, then the following conditions are equivalent:

(1) D is a Prüfer domain.
(2) $D(X) = D^K$.
(3) $D(X)$ is a Prüfer domain.
(4) D^K is a quotient ring of $D[X]$.
(5) Each prime ideal of $D(X)$ is a contraction of a prime ideal of D^K.
(6) Each prime ideal of $D(X)$ is an extension of a prime ideal of D.

Theorem (21.4) is a generalization of Arnold's theorem to Marot rings with Property A. The equivalence of conditions (1)–(4) of (21.4) is proved in [51]. In 1980 (5) and (6) of (21.4) were shown to be equivalent to (1)–(4), if the additively regular hypothesis i assumed [56]. Three years later Matsuda gave a proof of the equivalence of (5) and (6) with (1)–(4) for Marot rings, [81]. Theorem 21.2, when published, was new even for integral domains, and it too is from [56].

Section 22 Lemma 22.4 from [56] is a generalization of a similar theorem for integral domains given by Arnold and Brewer [14]. Special cases of (22.5) are given in [57] and [14]. The proof of 22.5 is in [56]. Matsuda [82] proved (22.7).

Chapter V

Chained Rings

A question, attributed to I. Kaplansky, is whether every chained ring is the homomorphic image of a valuation domain. Recently L. Fuchs and L. Salce have given an example in their book [FS] to show that this is not always true. In this chapter chained rings are discussed and instances are given where the answer to Kaplansky's question is positive.

23. Chained Rings

A ring R is *chained* if its set of ideals is linearly ordered by inclusion. A chained ring has few zero divisors and is therefore a Marot ring. Thus a chained ring is a Prüfer valuation ring (Theorem 7.7). It is easy to see that the converse does not hold. For an example, let T be a total quotient ring that is not chained and let

$P \in \operatorname{Spec} T$. Then T becomes a Prüfer valuation ring by defining $v(x) = 0$ if $x \in P$ and $v(x) = \infty$ if $x \in T \setminus P$.

Theorem 23.1 Let v be a valuation on a total quotient ring T with corresponding valuation ring V. If V is a chained ring, then $Z(T)$ is a maximal ideal of T and $Z(V) = \{x \in T : v(x) = \infty\} = Z(T)$. Moreover, there is a valuation $\bar{v}$ on the field $T/Z(T)$ such that v is given by $v(x) = \bar{v}(\bar{x})$, where $\bar{x}$ is the residue of x modulo $Z(T)$.

PROOF: Note that $T(V) = T$. For if $x \in T$ such that $v(x) < 0$, then there exists a regular $y \in V$ such that $v(x) + v(y) = 0$ (Theorem 7.9), and hence $x = r/y$ for some $r \in V$.

Since V is a chained ring, $Z(V)$ is a prime ideal of V, and $Z(V)T = Z(T)$ is a maximal ideal of T.

It is clear that $\{x \in T : v(x) = \infty\} \subseteq Z(V)$. Suppose that $x \in Z(V)$ such that $v(x) < \infty$. If $v(x) = 0$, then x does not belong to the unique maximal ideal of V; so x is a unit in V, a contradiction. Thus $0 < v(x) < \infty$. Use Theorem 7.9 to find a regular element r in V such that $v(x) = v(r)$. Thus $x = ry$ where $y \in Z(V)$ and $v(x) = v(r) + v(y)$. But then $v(y) = 0$, which we have already seen to be impossible. This proves that $Z(V) = \{x \in T : v(x) = \infty\}$. We still need to prove that $Z(V) = Z(T)$. If $z \in Z(T) = Z(V)T$, then $z = a/b$ where $a \in Z(V)$ and b is a regular element of V. Then $v(z) = \infty$, so $Z(T) = Z(V)$.

For the second statement of the theorem, let G be the value group of v and define $\bar{v} : T/Z(T) \to G \cup \{\infty\}$ by $\bar{v}(\bar{x}) = v(x)$. The first part of the theorem shows that $\bar{v}$ is well defined. The rest is clear.

If T is a chained total quotient ring, what type of valuation rings must T possess? The next theorem answers this question.

Theorem 23.2 Let T be a chained total quotient ring and let

V be a valuation subring of T such that $T(V) = T$. Then V is a chained ring if and only if $Z(T) \subseteq V$. Otherwise V has exactly two maximal ideals, namely, $Z(V)$ and a regular maximal ideal.

PROOF: Theorem 23.1 implies that if V is a chained ring, then $Z(T) \subset V$. For the converse, assume that $Z(T) \subseteq V$. Let $x, y \in V$ and suppose that $xT \subseteq yT$. Then $x = (a/s)y$ where $a, s \in V$ and s is regular. If a is regular, then $v(a/s) \geq 0$ or $v(s/a) \geq 0$. Hence $xV \subseteq yV$ or $yV \subseteq xV$. If a is a zero divisor, then $a/s \in Z(T) \subseteq V$, so $xV \subseteq yV$.

For the final statement of the theorem, assume that $Z(T) \not\subseteq V$. Since V has few zero divisors, V is a Prüfer valuation ring (Theorem 7.7) with a unique regular maximal ideal M (Theorem 6.5). Recall that $M = \{x \in T : v(x) > 0\}$. Let $a/s \in Z(T) \setminus V$, where $a \in Z(V)$ and s is a regular element of V. By Theorem 7.9 there exists a regular $r \in V$ such that $v(a) = v(r)$; that is, $a/r \in V \setminus M$. Clearly $a/r \in Z(V) = Z(T) \cap V$. Since T is a chained ring, $Z(T)$ is an ideal of T. Thus $Z(V)$ is an ideal of V. If $Z(V) \subset N$ where N is a maximal ideal of V, then N is regular, so $Z(V) \subset N = M$, a contradiction. This proves that M and $Z(V)$ are the only maximal ideals of V.

Each homomorphic image of a valuation domain is a chained ring. An example will be given later to show that there exists a chained total quotient ring T that admits a valuation ring V which is not chained and such that $T(V) = T$.

Corollary 23.3 Let V be a valuation ring such that $V \neq T(V)$. Assume that $T(V) = T$ is chained.

(1) If V is a chained ring, then every overring of V is chained.

(2) If V has exactly two maximal ideals, then every overring of V other than T has exactly two maximal ideals.

PROOF: (1): If $Z(T) \subset V$, then $Z(T)$ is contained in each overring of V.

(2): Let W be an overring of V such that $W \neq T$. Since W is a Prüfer valuation ring, it has a unique regular maximal ideal N. Let M be the unique regular maximal ideal of V. As in the integral domain case, each regular element of N is contained in M. But M and N are V-modules and V is a Marot ring; thus $N \subseteq M$. If W is a chained ring, then $Z(T) \subseteq N \subseteq M \subseteq V$, which implies that V is chained, a contradiction.

We turn our attention to semigroup rings. *All groups and semigroups are assumed to be commutative and will be written with additive notation. Furthermore, semigroups are assumed to have an identity element.* Thus each semigroup is a *monoid.* For a ring R and a monoid S, define $R[S]$ to be the set of all formal sums $\sum r_i g_i$ where $r_i \in R$ and $g_i \in G$, and only finitely many r_i are nonzero. Changing notation we write $r_1 X^{g_1} + \cdots + r_n X^{g_n}$ in place of $r_1 g_1 + \cdots + r_n g_n$. Define addition and multiplication in $R[S]$ the same as in polynomial rings. (Note: $X^{g_1} X^{g_2} = X^{g_1 + g_2}$.) Under these definitions $R[S]$ becomes a commutative ring with identity, called the *monoid ring* of R over S. This is a special type of semigroup ring. An element $r_1 X^{g_1} + \cdots + r_n X^{g_n}$ of $R[S]$ is in *canonical form* if each $r_i \neq 0$ and if $g_i \neq g_j$ for $i \neq j$. Each element of $R[S]$ may be put into canonical form. It is easy to see that two elements are equal if and only if they have the same canonical form. There is a natural embedding of R into $R[S]$ given by $r \mapsto rX^o$. We usually consider $R \subseteq R[S]$ and write r instead of rX^o. The function $\varepsilon : R[S] \to R$ defined by $\varepsilon(\sum r_i X^{g_i}) = \sum r_i$ is a ring homomorphism called the *augmentation map.* The kernel of this map is called the *augmentation ideal* of $R[S]$. We prove two main results about monoid rings. First we give a characterization of quasilocal monoid rings. Using this we then give necessary and sufficient conditions for a monoid ring to be chained.

If G is a group and $E \subseteq G$, denote the subgroup of G generated by E by $\langle E \rangle$.

Lemma 23.4 Let p be a prime number. If G is an abelian p-group and if $\mathcal{M}$ is a maximal ideal of the group ring $R[G]$, then $\mathcal{M} \cap R = M$ is a maximal ideal of R.

PROOF: Let $K = R[G]/\mathcal{M}$ and $\bar{R} = R/M$, and let $\bar{r} = r + M$ be a nonzero element in $\bar{R}$. Since $R/M \subseteq R[G]/\mathcal{M}$, there exists $\bar{y} = y + \mathcal{M} \in R[G]/\mathcal{M}$ such that $\bar{y}\bar{r} = \bar{1}$. Write $y = \sum_{i=1}^{s} r_i X^{g_i}$ and let H be the subgroup of G generated by the g_i's. But H is a finitely generated p-group, so H is finite, say $H = \{h_1, \ldots, h_t\}$. Rewrite y as $y = \sum_{i=1}^{t} r_i X^{h_i}$. Then for $j = 1, \ldots, t$ we have $yx^{h_j} = \sum_{i=1}^{t} r_{ji} X^{h_i}$. This leads to the system of equations $\sum_{i=1}^{t} (r_{ji} - \delta_{ji} y) X^{h_i} = 0$, where δ_{ji} is the Kronecker symbol. It follows that if d is the determinant of the matrix $(r_{ji} - \delta_{ji} y)$, then $d = 0$. Expanding d yields $y^t + a_1 y^{t-1} + \cdots + a_t = 0$, where $a_i \in R$. Multiply the last equation by r^{t-1}. We get $y(yr)^{t-1} = -[a_1(yr)^{t-1} + \cdots + r^{t-1} a_t]$. Let $y' = -[a_1 + a_2 r + \cdots + a_t r^{t-1}]$. We have equality of cosets $y(yr)^{t-1} + \mathcal{M} = -[a_1(yr)^{t-1} + \cdots + a_t r^{t-1}] + \mathcal{M}$. Since $yr + \mathcal{M} = 1 + \mathcal{M}, y + \mathcal{M} = -[a_1 + \cdots + r^{t-1} a_t] + \mathcal{M} = y' + \mathcal{M}$ with $y' \in R$. Thus $(y' + \mathcal{M})(r + \mathcal{M}) = 1 + \mathcal{M}$, so $y'r - 1 \in \mathcal{M} \cap R = M$; hence R/M is a field.

Lemma 23.5 Let p be a prime integer and let G be an abelian p-group. If $\mathcal{M}$ is maximal ideal of $R[G]$ such that $R/(\mathcal{M} \cap R)$ has characteristic p, then $\mathcal{M} = \varepsilon^{-1}(\mathcal{M} \cap R)$.

PROOF: Let $\sigma : R[G] \to R[G]/\mathcal{M}$ be the canonical homomorphism and let $\theta : R \to R/(\mathcal{M} \cap R)$ be its restriction. Let $x = \sum_{i=1}^{s} r_i X^{g_i} \in R[G]$ and choose t such that $p^t g_i = 0$ for $i = 1, \ldots, s$. Then $\sigma(1) = \sigma(X^0) = \sigma(X^{g_i})^{p^t}$, and so $0 = \sigma(1) - \sigma(X^{g_i})^{p^t} = [\sigma(1) - \sigma(X^{g_i})]^{p^t}$. Therefore $\sigma(1) = \sigma(X^{g_i})$ for each i. Consequently $\sigma(x) = \sum \theta(r_i)\sigma(X^{g_i}) = \sum \theta(r_i) = \theta(\varepsilon(x))$. Therefore $\mathcal{M} = \ker(\sigma) = \varepsilon^{-1}(\ker(\theta)) = \varepsilon^{-1}(\mathcal{M} \cap R)$.

A nonempty subset I of a semigroup S is an *ideal* of S if $I + S \subseteq I$. Assume that $g \in S$ and let $J = g + S$, then J is an ideal of S. If $J = S$, then $g + s = 0$ for some $s \in S$, and hence g has an inverse in S. Thus if S has no proper ideals, then S is a group. Note that I is a proper ideal of S if and only if $0 \notin I$.

Theorem 23.6 Let p be a prime number. If (R, M) is a quasilocal ring such that the characteristic of R/M is p and if G is an abelian p-group, then $R[G]$ is a quasilocal ring.

Conversely, if $(R[S], \mathcal{M})$ is a quasilocal ring, then R is quasilocal with maximal ideal $\mathcal{M} \cap R = M$, R/M has characteristic $p \neq 0$, and S is an abelian p-group.

PROOF: For the first part, assume that $\mathcal{M}$ and $\mathcal{N}$ are maximal ideals of $R[G]$. By Lemma 23.4, $\mathcal{M} \cap R = M = \mathcal{N} \cap R$, the unique maximal ideal of R. Lemma 23.5 implies that $\mathcal{M} = \varepsilon^{-1}(M) = \mathcal{N}$.

Conversely, assume that $(R[S], \mathcal{M})$ is a quasilocal ring. We prove that S is a group by showing that it contains no proper ideals. Assume that I is a proper ideal of S. We look for a contradiction. Define an equivalence relation $\sim$ on S by $g \sim h$ if and only if $g = h$ or $g, h \in I$. This relation is compatible with the binary operation on S. If $[g]$ is the equivalence class of g, let $S/I = \{[g] : g \in S\}$. Then S/I is a semigroup and there is a natural homomorphism ρ from S onto S/I. This may be extended to a ring epimorphism $\rho^* : R[S] \to R[S/I]$. Then $R[S/I]$ is a quasilocal ring. However if we denote the equivalence class of I by ∞, then it is easily seen that X^∞ is an idempotent element in $R[S/I]$ and X^∞ is different from $1 = 1X^{[0]}$ and $0 = 0X^{[0]}$. Thus $R[S/I]$ cannot be quasilocal. Therefore S is a group.

Let p be charR/M; that is, the characteristic of R/M, if char$R/M \neq 0$. If char$R/M = 0$, let $p = 1$. For a fixed $g \in S$ we claim that the order of g is a power of p; this will imply that $p > 1$ and S is a p-group. Suppose not. If g has finite order that is not a power of p, then there exists a prime integer $q \neq p$ such that S

contains an element of order q. If g has infinite order, let q be any prime number different from p. Let T be the group of all roots of unity in the algebraic closure of the field R/M. Define $\rho_o : \langle g \rangle \to T$ by $\rho_o(g) = u$ where u is a fixed primitive qth root of unity. By the restrictions placed on q, ρ_o is a homomorphism. Since T is a divisible group, it is an injective Z-module. Hence ρ_o may be extended to a homomorphism $\rho : S \to T$. If $H = \rho(S)$, then H is a (multiplicative) subgroup of the algebraic closure of R/M and we may consider the algebraic extension field $(R/M)[H] = (R/M)(H)$.

Let $\theta : R \to R/M$ be the natural epimorphism. Define a mapping $\gamma : R[S] \to (R/M)(H)$ by $\gamma(\sum r_i X^{g_i}) = \sum \theta(r_i)\rho(g_i)$. Then γ is an epimorphism, so its kernel $\mathcal{N}$ is a maximal ideal of $R[G]$. But $X^0 - X^g$ is a nonunit of $R[S]$, because if there exists some $f \in R[S]$ such that $(X^0 - X^g)f = 1$, then $\varepsilon((X^0 - X^g)f) = \varepsilon(1)$ implies $0 = 1$. Therefore $X^0 - X^g$ is in $\mathcal{M}$, while $\gamma(X^0 - X^g) = \theta(1)\rho(0) - \theta(1)\rho(g) = 1 - u \neq 0$. Hence $X^0 - X^g \notin \mathcal{N}$. This contradicts the fact that $R[G]$ is a quasilocal ring. Since $R \cong R[G]/\ker\varepsilon$, where ε is the augmentation map, R is a quasilocal ring. Also $\mathcal{M} \cap R = M$ is the unique maximal ideal of R (Lemma 23.4).

Before proceeding to the next theorem, we need to discuss some abelian group theory. Fix a prime integer p. Let $C(p^\infty)$ be the multiplicative group of all $p^{i^{th}}$ complex roots of 1, $i = 1, 2, \ldots$. Write $C(P^\infty)$ in additive notation. This group may be defined abstractly as the group generated by elements $c_1, c_2, c_3, \ldots$ such that $pc_1 = 0$, $pc_2 = c_1$, $pc_3 = c_2, \ldots$. Thus the order of c_i is p^i for $i = 1, 2, \ldots$. The nonzero proper subgroups of $C(p^\infty)$ are $\langle c_1 \rangle \subset \langle c_2 \rangle \subset \langle c_3 \rangle \subset \cdots$. A group G is *cocyclic* if it is isomorphic to some $\langle c_i \rangle, i = 1, 2, \ldots$; or to $C(p^\infty)$. Each cocyclic group is an abelian p-group.

Lemma 23.7 A p-group G is cocyclic if and only if G has the property that if $g, h \in G$, then there exists an integer m such that $g = mh$ or $h = mg$.

PROOF: ($\Rightarrow$): Without loss of generality assume that $G = C(p^\infty)$. If g and h are in G, then there is some $i < \infty$ such that $g, h \in \langle c_i \rangle$. Write $g = dac_i$ and $h = dbc_i$ where a, b, and d are integers, and a and b are relatively prime. Assume that p does not divide a. Choose integers u and v such that $1 = au + p^i v$. Thus $h = audbc_i + p^i vdbc_i = ubg$; so $m = ub$.

($\Leftarrow$): Choose $g \in G$ of order p and let H be a nonzero subgroup of G. If h is a nonzero element in H, then there is an integer m such that $g = mh$ or $h = mg$. If $g = mh$, then $g \in H$. On the other hand, if $h = mg$, then $\langle h \rangle \subseteq \langle g \rangle$, whence $\langle h \rangle = \langle g \rangle \subseteq H$. Hence g is contained in the intersection of all nonzero subgroups of G. Therefore $\langle g \rangle$ is the smallest subgroup of G.

Proceed by induction. Assume that G contains at most one subgroup of order p^n and that it is cyclic. We may as well assume that this subgroup is $\langle c_n \rangle$. Suppose that there are two subgroups H and K of G of order p^{n+1}, and that $h \in H \backslash \langle c_n \rangle$ and $k \in K \backslash \langle c_n \rangle$. Then h and k each have order p^{n+1} which forces H and K to be cyclic. Thus $ph = rc_n$ and $pk = sc_n$ for integers r and s. Then r (resp., s) is relatively prime to p. Otherwise the order of h (resp., k) is $< p^{n+1}$. Hence there exists integers r' and s' such that $rr' \equiv 1 (\operatorname{mod} p^n)$ and $ss' \equiv 1 (\operatorname{mod} p^n)$. Notice that r' and s' are relatively prime to p. Then $h' = r'h$ and $k' = s'k$ satisfy the relations $\langle h' \rangle = \langle h \rangle, \langle k' \rangle = \langle k \rangle$, $ph' = c_n$ and $pk' = c_n$. Hence $p(h' - k') = 0$ and thus $h' - k' = tc_n$ for some t. Consequently $h' = k' + tpk'$ and $k' = h' - tph'$. Therefore $H = K$. Thus G is the union of a finite or infinite chain of cyclic subgroups $H_1 \subset H_2 \subset H_3 \subset \cdots$, where each $H_i \cong \langle c_i \rangle$.

Lemma 23.8 If g and h are nonzero elements of a group G and if $1 - X^g$ divides $1 - X^h$ in the monoid ring $R[G]$, then $h = mg$ for some integer m.

PROOF: Assume that no such m exists. Write $1 - X^h = (1 - X^g)(a_1 X^{g_1} + \cdots + a_n X^{g_n})$ where $a_i \in R$, $g_i \in G$, and $g_i \neq g_j$ if $i \neq j$. Some $g_i = 0$ or $-g$, hence some $g_i \in \langle g \rangle$. On the other hand if all $g_i \in \langle g \rangle$, then $h \in \langle g \rangle$, so $h = mg$, contradicting the first sentence. Assume that $g_1, \ldots, g_t \in \langle g \rangle$ and $g_{t+1}, \ldots, g_n \notin \langle g \rangle$, where $1 \leq t \leq n - 1$. Clearly

$$\{0, g_i, g_i + g\}_{i=1}^{t} \cap \{h, g_i, g_i + g\}_{i=t+1}^{n} = \varnothing$$

Thus

$$\begin{aligned} 1 &= (1 - X^g)(a_1 X^{g_1} + \cdots + a_t X^{g_t}) \\ &\quad + (1 - X^g)(a_{t+1} X^{g_{t+1}} + \cdots + a_n X^{g_n}) + X^h \\ &= (1 - X^g)(a_1 X^{g_1} + \cdots + a_t X^{g_t}) \end{aligned}$$

This cannot hold, since the augmentation map ε sends 1 to 1 and $(1 - X^g)(a_1 X^{g_1} + \cdots + a_t X^{g_t})$ to 0. Therefore there exists some integer m such that $h = mg$.

Lemma 23.9 Let K be a field, $K[X]$ the polynomial ring over K, Z the additive group of integers, and $\varphi : Z \to \langle c_i \rangle$ be the canonical homomorphism. Then $\varphi^* : K[X] \to K[\langle c_i \rangle)]$ defined by $\varphi^*(\sum a_j X^j) = \sum a_j X^{\varphi(j)}$ is an epimorphism and $\ker \varphi^* = (1 - X^{p^i})$.

PROOF: Certainly $(1 - X^{p^i}) \subseteq \ker \varphi^*$. Let $f \in \ker \varphi^*$. If $f \neq 0$, then it must have at least two nonzero terms. Say that $f = a_k X^k + a_j X^j \in \ker \varphi^*$. Then $\varphi^*(f) = a_k X^{\varphi(k)} + a_j X^{\varphi(j)} = 0$; so $a_k = -a_j$ and $\varphi(k) = \varphi(j)$. Write $j = k + tp^i$; then $f = a_k X^g(1 - X^{p^i}) \in (1 - X^{p^i})$. Proceed by induction on the number of nonzero terms of f. Let $f = a_1 X^{n(1)} + \cdots + a_m X^{n(m)}$ where each $a_j \neq 0$, $m > 2$, and if $j \neq k$ then $n(j) \neq n(k)$. If $f \in \ker \varphi^*$, choose $k \neq j$ such that $\varphi(k) = \varphi(j)$. If $g = f - (a_k X^{n(k)} - a_k X^{n(j)})$, then $\varphi^*(g) = 0$

and the number of nonzero terms of g is $< m$. By induction $g \in (1 - X^{p^i})$, and hence $\ker \varphi^* = (1 - X^{p^i})$.

The final result of this section characterizes chained semigroup rings.

Theorem 23.10 Let S be a nonzero monoid and let R be a ring. Then $R[S]$ is chained if and only if R is a field of characteristic $p > 0$ and S is a cocyclic p-group.

PROOF: ($\Rightarrow$): By Theorem 23.6, S is an abelian p-group and R is quasilocal with maximal ideal M such that R/M has characteristic $p > 0$. Lemmas 23.7 and 23.8 imply that S is cocyclic. It remains to show that R is a field. Choose nonzero elements $a \in R$ and $g \in S$. One of a or $1 - X^g$ divides the other. But $a = (1 - X^g)r$ cannot occur because $a = \varepsilon(a) = \varepsilon((1 - X^g)r) = 0$. Thus $a(a_1X^{g_1} + \cdots + a_nX^{g_n}) = 1 - X^g$ where $g_i \neq g_j$ if $i \neq j$. There is exactly one i such that $g_i = 0$. Hence $aa_i = 1$.

($\Leftarrow$): Either $S = \langle c_i \rangle$ for $i < \infty$ or $S = C(p^\infty) = \cup_{i=1}^{\infty} \langle c_i \rangle$. Hence it suffices to prove that if $i < \infty$, then $R[\langle c_i \rangle]$ is chained. Let φ^* be the epimorphism defined in Lemma 23.9. Then $\ker \varphi^* = (X^{p^i} - 1) = (X - 1)^{p^i}$. Thus $R[X]/(X - 1)^{p^i} = R[\langle c_i \rangle]$. The proper ideals of $R[X]/(X - 1)^{p^i}$ are

$$(0)/(X - 1)^{p^i} \subset (X - 1)^{p^{i-1}}/(X - 1)^{p^i} \subset \cdots \subset (X - 1)/(X - 1)^{p^i}$$

24. Homomorphic Images of Valuation Domains

This section gives the known situations for which the answer to the Kaplansky question is positive: Theorems 24.3 and 24.4. The first results show how to reduce the problem to a question about total quotient rings.

Theorem 24.1 Let R be a Prüfer ring. Then R is a homomorphic image of a Prüfer domain if and only if its total quotient ring is a homomorphic image of a Prüfer domain.

PROOF: Let R be a Prüfer ring, let D be a Prüfer domain, and let $\theta : D \to R$ be an epimorphism. If $N = \{x \in D : \theta(x) \text{ is regular}\}$, then N is a multiplicatively closed subset of D and hence D_N is a Prüfer domain. Extend θ to D_N by defining $\theta^*(x/y) = \theta(x)/\theta(y)$. The range of θ^* is $T(R)$.

Conversely, assume that there is a Prüfer domain K and a homomorphism φ from K onto $T(R)$. Without loss of generality we may assume that $\varphi(x)$ is a unit in $T(R)$ if and only if x is a unit in K. (Replace K by the Prüfer domain K_N where $N = \{x \in K : \varphi(x) \text{ is a unit in } T(R)\}$. Then map $K_N \to T(R)$ by sending x/y to $\varphi(x)/\varphi(y)$.) Define $D = \{x \in K : \varphi(x) \in R\}$ and let θ be the restriction of φ to D. If $S = \{d \in D : \theta(d) \text{ is regular}\}$, then S is a multiplicatively closed subset, $D_S \subseteq K$, and $\varphi(D_S) = T(R)$. If $k \in K$, then $\varphi(k) = \theta(a)/\theta(b)$ where $a, b \in D$ with $b \in S$. Thus $k - a/b = t \in \ker \varphi$. By the definition of D we have $\ker \varphi \subseteq D$, forcing $k = (a + bt)/b \in D_S$. Therefore $D_S = K$.

We prove that D is a Prüfer domain by showing that if a_1 and a_2 are elements of D, not both zero, then $I = (a_1, a_2)$ is an invertible ideal [G, p. 276]. The K-ideal IK is invertible. Choose $t_1, t_2 \in (IK)^{-1}$ such that $a_1t_1 + a_2t_2 = 1$. Since $D_S = K$ there exists some $s \in S$ such that $st_1 = c_1$ and $st_2 = c_2$ belong to I^{-1}; hence $a_1c_1 + a_2c_2 = s$. Let $J = (a_1, a_2, a_1c_1, a_2c_1, a_1c_2, a_2c_2)$. Then $\theta(J)$ is a regular ideal of R, because $\theta(a_1c_1 + a_2c_2) = \theta(s) = \varphi(s)$ is a unit of $T(R)$ contained in R. For the moment assume that J is an invertible ideal of D. Then there exists $e_i \in J^{-1}$ such that

$$a_1e_1 + a_2e_2 + a_1c_1e_3 + a_2c_1e_4 + a_1c_2e_5 + a_2c_2e_6 = 1$$

Clearly $e_1, e_2, c_1e_3, c_1e_4, c_2e_5, c_2e_6 \in I^{-1}$. Thus I is an invertible ideal.

To complete the proof we need to show the following: If $B = (b_1, \ldots, b_n)$ is a finitely generated ideal of D such that $\theta(B)$ is a regular ideal of R, then B is an invertible ideal of D. Assume that B is as above. Since R is a Prüfer ring, we may choose $t_i \in K$ such that $\varphi(t_i) \in \theta(B)^{-1}$ and $\sum \varphi(t_i)\theta(b_i) = 1$. Each $\varphi(t_i)\theta(b_i) \in R$, thus each $t_i b_i \in D$, and $\varphi(\sum t_i b_i) = 1$ implies that $d = \sum t_i b_i$ is a unit of K contained in D. Since $\varphi(d^{-1}) = 1, d^{-1}$ is also in D; that is, d is a unit in D. Therefore $BB^{-1} = D$.

Corollary 24.2 Let R be a chained ring. Then R is a homomorphic image of a valuation domain if and only if $T(R)$ is the homomorphic image of a valuation domain.

PROOF: ($\Leftarrow$): By Theorem 24.1, there is a Prüfer domain D and an epimorphism $\varphi : D \to R$. Let M be the unique maximal ideal of R and let N be the maximal ideal of D that maps onto M under φ. Extend φ to $\varphi^* : D_N \to R$ by defining $\varphi^*(x/y) = \varphi(x)/\varphi(y)$. Thus R is the homomorphic image of a valuation domain.

($\Rightarrow$): If R is the homomorphic image of the valuation domain V, then as in the proof of Theorem 24.1, $T(R)$ is the homomorphic image of V_S where S is a multiplicatively closed set.

Recall the Cohen structure theorem for complete local rings: Let (R, M) be a complete local (Noetherian) ring such that M has a minimal basis of n elements. Then R is a homomorphic image of a power series ring $D[[X_1, \ldots, X_n]]$. If R and R/M have the same characteristic, then $D = R/M$. If $\operatorname{char} R \neq \operatorname{char} R/M$, then $D = V$ is a discrete rank one valuation domain of characteristic 0; and if P is the maximal ideal of V, then $R/M \cong V/P$ has characteristic $p \neq 0$. Some versions and proofs of this result may be found in [N, p. 106] and [ZSII, p. 307]. For the exact statement that we have given here, consult Cohen's original paper.

Theorem 24.3 If R is a chained ring with a Noetherian total quotient ring, then R is a homomorphic image of a valuation domain.

PROOF: In view of Corollary 24.2, it suffices to show that $T(R)$ is a homomorphic image of a valuation domain. Denote $T(R)$ by T. Then T is a Noetherian chained ring and $Z(T) \supseteq N(T)$ are prime ideals of T. We show that $N(T) = Z(T)$. If $N(T) = (0)$, then T is an integral domain, so $Z(T) = N(T)$. Suppose that $Z(T) \supset N(T) \supset (0)$. Then $\cap_{i=1}^{\infty} Z(T)^i \supseteq N(T) \supset (0)$. But this is impossible in a Noetherian local ring. Thus T is a special primary ideal ring whose set of proper ideals are $N(T) \supset N(T)^2 \supset \cdots \supset N(T)^s = (0)$. Let $N(T) = (x)$. Clearly T is a complete local ring.

If $\operatorname{char} T = \operatorname{char} T/N(T)$, then T is the homomorphic image of the valuation domain $(T/N(T))[[X]]$. Assume that $\operatorname{char} T \neq \operatorname{char} T/N(T)$; then T is the homomorphic image of $V[[X]]$ where V is a rank one discrete valuation domain with principal maximal ideal $P = (a)$.

Since $V[[X]]$ is a two-dimensional Noetherian ring [G, p. 357] with unique maximal ideal $\mathcal{M}$ generated by a and X [N, p. 49], $V[[X]]$ is a regular local ring [ZSII, p. 301]. Let $\varphi : V[[X]] \rightarrow T$ be the epimorphism. Either a or X maps onto a generator of $N(T)$. If x is a generator of $N(T)$, then we may assume that either $\varphi(a) = x$ or $\varphi(X) = x$. If $\varphi(a) = x$, then $\varphi(X) = ux^i$ where u is a unit in T. Choose g such that $\varphi(g) = u^{-1}$; then $a^i - gX \in \ker \varphi$. It follows that

$$V[[X]]/(a^i - gX) \xrightarrow{\beta} V[[X]]/\ker \varphi \cong T$$

where β is the natural homomorphism. Since $(a, a^i - gX) = (a, X) = \mathcal{M}$, $V[[X]]/(a^i - gX)$ is a one-dimensional regular local ring [K, pp. 120 and 117], that is, T is the homomorphic image of a valuation domain. The case where $\varphi(X) = x$ is proved in a similar manner.

The rest of this section is devoted to proving that a chained monoid ring is a homomorphic image of a valuation domain. Before doing this, some material concerning general graded rings is needed. We give a brief outline of the pertinent facts. For more detail and for the proofs of the results listed below, see Northcott's book [No, pp. 112–117]. Let Γ be a commutative cancellative monoid (additively written). A ring R is *graded by* Γ if there is a family $\{R_\gamma\}_{\gamma\in\Gamma}$ of subgroups of the additive group of R such that

(1) $R = \sum_{\gamma\in\Gamma} R_\gamma$ (direct);
(2) $R_{\gamma_1}R_{\gamma_2} \subseteq R_{\gamma_1+\gamma_2}$, for all $\gamma_1, \gamma_2 \in \Gamma$.

An element of $R_{\gamma_1}R_{\gamma_2}$ is a finite sum of elements of the form rs where $r \in R_{\gamma_1}$ and $s \in R_{\gamma_2}$. An element in R_γ is said to be *homogeneous of degree* γ. Clearly 0 in R is homogeneous of degree γ, for each γ. If A is an R-module and if R is graded by Γ, define A to be a *graded* Γ*-module* if there is a family of subgroups $\{A_\gamma\}_{\gamma\in\Gamma}$ of A such that

(1′) $A = \sum_{\gamma\in\Gamma} A_\gamma$ (direct);
(2′) $R_{\gamma_1}A_{\gamma_2} \subseteq A_{\gamma_1+\gamma_2}$, for all $\gamma_1, \gamma_2 \in \Gamma$.

Again the elements of A_γ are said to be *homogeneous of degree* γ. Each nonzero element of the graded ring R (resp., module A) has a unique representation of the form $\sum_{i=1}^{n} x_{\gamma_i}$, where x_{γ_i} is a nonzero element of R_{γ_i} (resp., A_{γ_i}). The identity element 1 of R is in R_0 and R_0 is a subring of R. Hence condition (2′) implies that each A_γ is an R_0-module. A submodule B of A is *homogeneous* if B is generated as an R-module by homogeneous elements. This is equivalent to : $B = \sum_{\gamma\in\Gamma}(B \cap A_\gamma)$. If $A = R$, we have the concept of a *homogeneous ideal* of R. Assume that B is a homogeneous submodule of A. Let $\varphi : A \to A/B$ be the canonical homomorphism. If $\varphi(A_\gamma)$ is denoted by $(A/B)_\gamma$, then A/B becomes a graded R-module with Γ as the grading monoid and $\{(A/B)_\gamma\}_{\gamma\in\Gamma}$ as the defining subgroups of A/B. This is called the *factor grading* of A/B.

Theorem 24.4 If $R[S]$ is a chained monoid ring, then $R[S]$ is the homomorphic image of a valuation domain.

PROOF: By Theorem 23.10, R is a field of characteristic $p > 0$ and S is a cocyclic p-group. We consider the case where $S = \langle c_i \rangle$, for $i < \infty$. By Lemma 23.9, $R[X]/(1 - X^{p^i}) \cong R[S]$. From $(1 - X^{p^i}) = (1 - X)^{p^i}$ it follows that $R[X]_{(1-X)}/(1 - X)^{p^i}_{(1-X)} \cong R[X]/(1 - X)^{p^i}$. Thus we have an epimorphism

$$R[X]_{(1-X)} \longrightarrow R[X]/(1 - X)^{p^i} \longrightarrow R[S]$$

where the first term is a valuation domain.

It remains to treat the case where $S = C(p^\infty)$. Choose generators $c_1, c_2, c_3, \ldots$ of S such that $pc_1 = 0, pc_2 = c_1, pc_3 = c_2, \ldots$. Then $R[S] = R[c_1, c_2, \ldots]$. Let $D = R[X_1, X_2, \ldots]/I$ where $\{X_1\}_{i=1}^{\infty}$ are indeterminates over R and let $I = (X_2^p - X_1, X_3^p - X_2, \ldots)$. We show that I is a prime ideal of $R[X_1, X_2, \ldots]$ by proving that for each n, $J = (X_2^p - X_1, \ldots, X_{n+1}^p - X_n)$ is a prime ideal of $R[X_1, \ldots, X_{n+1}]$. Fix an integer n and let $\mathcal{M}$ be the maximal ideal $(X_1, \ldots, X_{n+1})$ of $R[X_1, \ldots, X_{n+1}]$. Then $J \subseteq \mathcal{M}$, and we need only show that $J_{\mathcal{M}}$ is a prime ideal in $R[X_1, \ldots, X_{n+1}]_{\mathcal{M}}$. Since $\{X_2^p - X_1, \ldots, X_{n+1}^p - X_n, X_{n+1}\}$ is a minimal basis for the maximal ideal of the regular local domain $R[X_1, \ldots, X_{n+1}]_{\mathcal{M}}$, $R[X_1, \ldots, X_{n+1}]_{\mathcal{M}}/J_{\mathcal{M}}$ is also a regular local domain [ZSII, p. 303]. Thus $D = R[X_1, X_2, \ldots]/I$ is an integral domain.

Let φ be the homomorphism from $R[X_1, X_2, \ldots]$ onto $R[S]$ such that $\varphi(X_i) = X^{c_i}$ and φ leaves elements of R fixed.

Since $I \subseteq \ker \varphi$,

$$\begin{array}{ccccc} R[X_1, X_2, \ldots] & \xrightarrow{\pi} & D & \xrightarrow{\beta} & R[X_1, X_2, \ldots]/\ker\varphi \\ \downarrow \varphi & & & \swarrow \gamma & \\ R[S] & & & & \end{array}$$

where π is canonical, $\beta(f+I) = f+\ker\varphi$ and $\gamma(f+\ker\varphi) = \varphi(f)$. Hence there is an epimorphism $\psi(=\gamma\beta)$ such that $\psi : D \to R[S]$. Write $R[X_1, X_2, \ldots] = R[Y_1, Y_2, \ldots]$ where $Y_i = 1 - X_i$. Since the characteristic of R is p, $I = (Y_2^p - Y_1, Y_3^p - Y_2, \ldots)$.

Let Γ be the submonoid of the additive group of rational numbers of the form j/p^i, where i and j are nonnegative integers. Make $R[Y_1, Y_2, \ldots]$ into a Γ-graded ring by defining $\deg Y_i = 1/p^i$ and $\deg(r) = 0$ for a nonzero $r \in R$. The ideal I is homogeneous with respect to this grading. Thus $D = R[Y_1, Y_2, \ldots]/I$ has a factor grading with respect to Γ. Write $D = \sum_{\gamma\in\Gamma} D_\gamma$ (direct). A typical nonzero element of D has a homogeneous decomposition of the form $x = d_{\gamma_1} + \cdots + d_{\gamma_n}$, where $\gamma_i \neq \gamma_j$ if $i \neq j$, and $\gamma_1 = \min\{\gamma_i\}$. Define a function $v : D \to \Gamma \cup \{\infty\}$ by

$$v(x) = \begin{cases} \gamma_1 & \text{if } x \neq 0 \\ \infty & \text{if } x = 0 \end{cases}$$

If Γ_1 is the image of $D\backslash\{0\}$ under v, then v maps D onto $\Gamma_1 \cup \{\infty\}$, and for all $x, y \in D$:

(i) $v(xy) = v(x) + v(y)$.
(ii) $v(x + y) \geq \min\{v(x), v(y)\}$.
(iii) $v(1) = 0$.

Let G be the subgroup of the rationals generated by Γ_1. Then v extends to $T(D)$, the quotient field of D, by defining $v(x/y) = v(x) - v(y)$. Furthermore v is a valuation on $T(D)$ with value group G. Let V be the associated valuation ring of v.

Suppose that T is the multiplicatively closed subset of D consisting of all $t \in D$ such that $v(t) = 0$. We show that $D_T = V$. Since V is a rank one valuation domain (because its value group is contained in the reals), it is a maximal proper subring of $T(D)$. Hence it suffices to prove that $V \subseteq D_T \subset T(D)$. Clearly D_T is properly contained in $T(D)$. Let y_i be the image of Y_i in D. Then $y_2^p = y_1, y_3^p = y_2, \ldots$; and hence each of the elements $y_1, \ldots, y_i$

may be written in terms of y_i. Consequently for a nonzero $x \in D$ there is some integer i such that $x = y_i^m(a_o + a_1 y_i + \cdots + a_n y_i^n)$, where $a_j \in R$ and $a_o \neq 0$. Furthermore $v(x) = v(y_i^m) = m/p^i$. If x and y are nonzero elements of D such that $x/y \in V$, then there exists an i such that

$$\frac{x}{y} = \frac{y_i^m(a_o + \cdots + a_n y_i^n)}{y_i^s(b_o + \cdots + b_r y_i^r)}$$

where $a_o \neq 0 \neq b_0$. Then $0 \leq v(x/y) = v(x) - v(y) = (m - s)/p^i$, so $m \geq s$. Thus

$$\frac{x}{y} = \frac{y_i^{m-s}(a_o + \cdots + a_n y_i^n)}{b_o + \cdots + b_r y_i^r}$$

and $v(b_o + \cdots + b_r y_i^r) = 0$. Therefore $x/y \in D_T$.

The epimorphism $\psi : D \to R[S]$ that we established earlier may be extended to an epimorphism $\psi^* : D_T \to R[S]_{\psi(T)}$. If $t \in T$, then $t = b_o + b_1 y_i + \cdots + b_n y_i^n$ where $b_o \neq 0$. Then $\psi(t) = b_o + b_1(1 - X^{c_i}) + \cdots + b_n(1 - X^{c_i})^n$. But $1 - X^{c_i} \in \ker \varepsilon \subseteq \mathcal{P}$, where ε is the augmentation map and $\mathcal{P}$ is the unique maximal ideal of $R[S]$. Therefore $\psi(t)$ is a unit of $R[S]$. Thus $\psi^* : D_T \to R[S]_{\psi(T)} = R[S]$ and we are finished.

Necessary and sufficient conditions for Kaplansky's question to have an affirmative answer are not known.

Notes

Section 23 Froeschl [34] proved (23.2) and (23.3). Gulliksen, Ribenboim, and Viswanathan [47] proved (23.4), (23.5) and the first part of (23.6). Their paper also contains a proof of the second part of (23.6), when S is assumed to be a group. Hardy and

Shores [48] completed the proof of (23.6) by removing the hypothesis that S need be a group. Lemmas 23.7 and 23.8 are from Ohm and Vicknair [89]. Theorem 23.10 appears in Gilmer [36]. The necessary conditions of (23.10), for $R[S]$ to be a chained ring, is due to Kozhukhov [61]; however the proof presented here is from [89].

Section 24 This section presents the known affirmative answers to the Kaplansky question. Theorem 24.1 is by Boisen and Larsen [17]. Theorem 24.3 was noted first by Hungerford [58] in 1968 and later by McLean [72] in 1973. Both proofs depend on the Cohen structure theorem [21]. Theorem 24.4 is proved in [89]. Paper [102] by Shores is closely associated with the ideas in this chapter. For each chained ring he constructs an associated semigroup. These semigroups are characterized. He then shows how to associate a unique "value group" to each chained ring.

Chapter VI

Constructions and Examples

Properties of the idealization of a ring and a module, and of $A+B$ rings are developed. These constructions are important when one wants to construct a ring (with zero divisors) that has certain predetermined properties. The last section consists of examples that have been mentioned throughout the rest of the text.

25. Idealization

For this section the following notation is fixed: R is a ring, B is an R-module, $Z(B)$ is the set of zero divisors on B, and $S = R \setminus \{Z(R) \cup Z(B)\}$.

Consider $R(+)B = \{(r,b) : r \in R, b \in B\}$ and let (r,b) and (s,c) be two elements of $R(+)B$. Define:

(a) $(r,b) = (s,c)$ if $r = s$ and $b = c$.

(b) $(r,b) + (s,c) = (r + s, b + c)$.
(c) $(r,b)(s,c) = (rs, rc + sb)$.

Under these definitions $R(+)B$ becomes a commutative ring with identity. Call this ring the *idealization* of B in R.

The ring R can be embedded into the ring $R(+)B$ via the map that takes r to (r, θ). Throughout we use θ as both the zero element of B and the zero submodule of B. The set $0(+)B$ is an ideal of $R(+)B$, giving rise to the name idealization. Notice that if $B \neq \theta$, then $0(+)B$ is nilpotent of index 2. In this case $R(+)B$ is not reduced. The structure of $0(+)B$ as an R-module is essentially the same as the structure of $0(+)B$ as an $R(+)B$-module.

The purpose of this section is to develop the basic properties of $R(+)B$. These results will be used extensively in Section 27, when constructing examples. We begin by describing the ideals of $R(+)B$.

Theorem 25.1 Let R and B be as above.

(1) Let J be an ideal of $R(+)B$. If $I = \{r \in R : (r,b) \in J, \text{ for some } b \in B\}$ and if $C = \{c \in B : (r,c) \in J, \text{ for some } r \in R\}$, then I is an ideal of R, C is an R-submodule of B, $IB \subseteq C$ and $J = I(+)C$.
(2) If I is an ideal of R and if C is an R-submodule of B such that $IB \subseteq C$, then $I(+)C$ is an ideal of $R(+)B$.
(3) The ideal J of $R(+)B$ is prime if and only if $J = P(+)B$ where P is a prime ideal of R. Hence $\dim R = \dim R(+)B$.
(4) Let $I(+)C$ and $I'(+)C'$ be ideals of $R(+)B$. Then:
 (a) $(I(+)C) \cap (I'(+)C') = (I \cap I')(+)(C \cap C')$;
 (b) $(I(+)C)(I'(+)C') = II'(+)(IC' + I'C)$.
(5) If $I(+)C$ is an ideal of $R(+)B$, then $\operatorname{Rad}(I(+)C) = (\operatorname{Rad} I)(+)B$.
(6) A necessary and sufficient condition for (r,b) to be a unit in $R(+)B$ is that r be a unit in R.

PROOF: The proofs of (1) and (2) are routine.

(3): Assume that $J = P(+)C$ is a prime ideal of $R(+)B$. Suppose that $x, y \in R$ such that $xy \in P$. For any $c \in C, (x,c)(y,c) = (xy, xc + yc) \in J$. Thus (x,c) or (y,c) is in $P(+)C$; that is, x or y is in P. Now suppose that $C \subset B$ and choose $b \in B \setminus C$ and $p \in P$. Since $PB \subseteq C$, $(p,b)^2 = (p^2, pb + pb) \in P(+)C = J$. But $(p,b) \notin J$, contradicting the hypothesis that J is prime. We conclude that $B = C$.

The converse is easy.

(4): We prove part (b). That $(I(+)C)(I'(+)C') \subseteq II'(+)(IC' + I'C)$ is clear. Choose $(r,b) \in II'(+)(IC' + I'C)$ and write $(r,b) = (\sum_{i=1}^n x_i y_i, \sum_{j=1}^m (s_j a'_j + s'_j a_j))$. Rewrite as follows:

$$\sum_{i=1}^{p}(x_i y_i, s_i a'_i + s'_i a_i) + \sum_{i=p+1}^{\max(m,n)} (r_i, b_i)$$

where $p = \min(m,n)$ and according to whether m or n is larger, either $r_i = 0$ and $b_i = s_i a'_i + s'_i a_i$, or $r_i = x_i y_i$ and $b_i = \theta$; for $i > p$.

For $i = 1, \ldots, p$,

$$\begin{aligned}&(x_i y_i, s_i a'_i + s'_i a_i) \\ &\qquad = (x_i, \theta)(y_i, \theta) + (s_i, a_i)(s'_i, a'_i) - (s_i, \theta)(s'_i, \theta)\end{aligned}$$

If $n > m$, then $(x_i y_i, \theta) = (x_i, \theta)(y_i \theta)$, $i = p+1, \ldots, n$. If $m > n$, then $(0, s_i a'_i + s'_i a_i) = (s_i, a_i)(s'_i, a'_i) - (s_i, \theta)(s'_i, \theta)$; $i = p+1, \ldots, m$.

We have realized (r,b) as a finite sum of products from $I(+)C$ and $I'(+)C'$.

(5): Suppose that $(r,b) \in \operatorname{Rad}(I(+)C)$ and that $(r,b)^n \in I(+)C$. Then $(r,b)^n = (r^n, b')$ where $r^n \in I$ and $b' \in B$. Thus $\operatorname{Rad}(I(+)C) \subseteq (\operatorname{Rad} I)(+)B$. Let $(r,b) \in (\operatorname{Rad} I)(+)B$ and choose n such that $r^n \in I$. Consider $(r,b)^{n+1} = (r^{n+1}, (n+1)r^n b)$. Since $IB \subseteq C$, $(n+1)r^n b \in C$. Therefore $(r,b)^{n+1} \in I(+)C$.

(6): Clear.

A submodule C of B is *primary* if for all $r \in R$ and $x \in B$, $rx \in C$ and $x \notin C$ imply that $r^n B \subseteq C$ for some positive integer n. If C is a submodule of B, then by the *radical* of C we mean $\operatorname{Rad}\{r \in R : rB \subseteq C\}$. Write $\operatorname{Rad} C$ for the radical of C. If C is a primary submodule of B, then $\operatorname{Rad} C$ is a prime ideal of R [LM, p. 41].

Theorem 25.2 The ideal $I(+)C$ of $R(+)B$ is primary if and only if I is a primary ideal of R, C is a primary submodule of B, and either $\operatorname{Rad} I = \operatorname{Rad} C$ or $B = C$.

PROOF: ($\Rightarrow$): Easy calculations show that I is a primary ideal of R and that C is a primary submodule of B. To show that the radicals are equal when $B \neq C$ observe that for $x \in \operatorname{Rad} I$, there exists a positive integer n such that $x^n B \subseteq C$—that is, $x \in \operatorname{Rad} C$. On the other hand, let $x \in \operatorname{Rad} C$ and $b \in B \setminus C$; then $(x^m, \theta)(0, b) \in I(+)C$ for some m. Since $I(+)C$ is primary, $(x^m, \theta)^n \in I(+)C$. This says that $x^{mn} \in I$.

($\Leftarrow$): If $C = B$ the result is clear. Consider $C \neq B$ and $\operatorname{Rad} I = \operatorname{Rad} C$. Assume that $(r, b)(s, c) \in I(+)C$ and $(s, c) \notin I(+)C$. Either $s \notin I$, or $s \in I$ and $c \notin C$. If $s \notin I$, then $r^n \in I$ for a positive integer n. Thus $(r, b)^{n+1} = (r^{n+1}, (n+1)r^n b) \in I(+)C$. If $s \in I$ and $c \notin C$, then $IB \subseteq C$ implies that $rc \in C$. Thus $r^n B \subseteq C$ and hence $r \in \operatorname{Rad} C = \operatorname{Rad} I$. Hence $r^m \in I$ for some m. Therefore for a large enough integer $k, (r, b)^k \in I(+)C$.

In order to discuss quotient ring formation, total quotient rings, integral closure, etc., with respect to $R(+)B$, we need to describe the set of zero divisors of $R(+)B$. The next theorem does this.

Theorem 25.3 Let R be a ring and let B be an R-module. Then $(r, b) \in Z(R(+)B)$ if and only if $r \in Z(R) \cup Z(B)$.

PROOF: Let $(r,b) \in Z(R(+)B)$. We may as well assume that $r \neq 0$. Choose a nonzero $(s,c) \in R(+)B$ such that $(r,b)(s,c) = (rs, sb+rc) = (0,\theta)$. If $s = 0$, then $r \in Z(B)$; and if $s \neq 0$, then $r \in Z(R)$.

Conversely, let $(r,b) \in R(+)B$ and assume that $r \in Z(R) \cup Z(B)$. If $r \in Z(R)$, choose a nonzero $s \in R$ such that $rs = 0$. If $s \in \operatorname{Ann} B$, then $(r,b)(s,\theta) = (0,\theta)$. If $s \notin \operatorname{Ann} B$, choose an element $c \in B$ such that $sc \neq \theta$. Then $(r,b)(0,sc) = (0,\theta)$. Finally, if $r \in Z(B)$, then $rc = \theta$ for some nonzero $c \in B$. Therefore $(r,b)(0,c) = (0,\theta)$.

Lemma 25.4 Let $\mathcal{N}$ be a saturated multiplicatively closed subset of R. Then:

(1) $\mathcal{N}(+)\theta$ is multiplicatively closed in $R(+)B$ and its saturation is $\mathcal{N}(+)B$.
(2) $(R(+)B)_{\mathcal{N}(+)\theta} \cong R_{\mathcal{N}}(+)B_{\mathcal{N}}$.

PROOF: (1): Clear.

(2): Define $h : (R(+)B)_{\mathcal{N}}(+)\theta \to R_{\mathcal{N}}(+)B_{\mathcal{N}}$ by letting h send $(r,a)/(s,\theta)$ to $(r/s, a/s)$. Then h is the required isomorphism.

Employing the fact that S is a regular multiplicatively closed subset of R we have the following result.

Corollary 25.5 With the notation adopted at the beginning of this section, the following statements hold:

(1) $T(R(+)B) \cong R_S(+)B_S$.
(2) If $P \in \operatorname{Spec}(R)$, then $(R(+)B)_{P(+)B} \cong R_P(+)B_P$.
(3) If $Z(B) \subseteq Z(R)$, then $T(R(+)B) \cong T(R)(+)B_S$.

Theorem 25.6 If R' is the integral closure of R in $T(R)$, then the integral closure of $R(+)B$ in $T(R(+)B)$ is $(R' \cap R_S)(+)B_S$.

PROOF: Since S is a regular multiplicatively closed set in R, $R' \cap R_S \subseteq T(R)$. It is sufficient to show that $(r,b) \in R_S(+)B_S$ is integral over $R(+)B$ if and only if $r \in R' \cap R_S$. First we assume that (r,b) is integral over $R(+)B$; then there exists a polynomial $f(X) = X^n + (r_{n-1}, b_{n-1})X^{n-1} + \cdots + (r_0, b_0)$ such that $f((r,b)) = (0,\theta)$. But $f((r,b)) = (r^n + r_{n-1}r^{n-1} + \cdots + r_0, c)$, where $c \in B$. Thus $r \in R' \cap R_S$. For the converse assume that $(r,b) \in (R' \cap R_S)(+)B_S$. Then (r,θ) is integral over $R(+)B$, and since $(0,b)^2 = (0,\theta)$, $(0,b)$ is also integral over $R(+)B$. Therefore $(r,\theta) + (0,b) = (r,b)$ is integral over $R(+)B$.

Corollary 25.7 If R is an integrally closed ring, then $R(+)B_S$ is the integral closure of $R(+)B$ in $T(R(+)B)$.

The converse of Corollary 25.7 fails. Let R be a ring that is not integrally closed. Choose B to be the direct sum of R/P_α, P_α ranging over all prime ideals of R. Then $Z(B)$ is the set of nonunits of R. By Theorem 25.1(6), $R(+)B_S = T(R(+)B)$. Thus $R(+)B_S$ is integrally closed. However a partial converse of Corollary 25.7 does hold.

Corollary 25.8 If $Z(B) \subseteq Z(R)$, then $R(+)B_S$ is integrally closed if and only if R is integrally closed.

We are ready to consider the problem of when the idealization of R with B is a Prüfer ring. Necessary conditions for $R(+)B$ to be a Prüfer ring are that $B = B_S$ and R be integrally closed in R_S. To determine when $R(+)B_S$ is Prüfer we need to know what the regular ideals look like and how to compute the inverse of an ideal of $R(+)B_S$.

Theorem 25.9 An ideal $I(+)C$ of $R(+)B_S$ is a regular ideal if and only if $I \cap S \neq \emptyset$ and $C = B_S$. If $I(+)B_S$ is a regular ideal of $R(+)B_S$ and a set $\mathcal{A}$ generates I as an ideal of R, then the set $\{(a,\theta) : a \in \mathcal{A}\}$ generates the ideal $I(+)B_S$.

PROOF: Assume that $I \cap S \neq \emptyset$ and $C = B_S$. Then $I(+)B_S$ contains regular elements. Conversely, since $I(+)C$ is a regular ideal, there exists $r \in I \cap S$. Let $b \in B_S$; then $r^{-1}b \in B_S$. Thus $(r,\theta)(0,r^{-1}b) = (0,b) \in I(+)C$. Consequently $C = B_S$.

For the second part, assume that $I(+)B_S$ is a regular ideal $R(+)B_S$. If $x \in I$, then clearly (x,θ) is contained in the ideal generated by $\{(a,\theta) : a \in \mathcal{A}\}$. If $b \in B_S$, choose $t \in I \cap S$. Then $t^{-1}b \in B_S$ and $(0,b) = (t,\theta)(0,t^{-1}b)$. Thus $(0,b)$ is also contained in the ideal generated by $\{(a,\theta) : a \in \mathcal{A}\}$. This completes the proof.

Theorem 25.10 If $I(+)C$ is an ideal of $R(+)B_S$, then $(I(+)C)^{-1} = (I^{-1} \cap R_S)(+)B_S$. Furthermore, if $I(+)B_S$ is a regular ideal, then $I^{-1} \cap R_S = I^{-1}$.

PROOF: If $(r,m) \in (I(+)C)^{-1}$, then $(r,m) \in R_S(+)B_S$ such that $(r,m)(a,b) \in R(+)B_S$ for all $(a,b) \in I(+)C$. Thus $rI \subseteq R$; that is, $r \in I^{-1} \cap R_S$, and hence $(r,m) \in (I^{-1} \cap R_S)(+)B_S$. The other inclusion is obvious.

If $I(+)B_S$ is a regular ideal, then $I \cap S \neq \emptyset$. Let $s \in I \cap S$ and $r \in I^{-1}$, then $rs = t \in R$. It follows that $r = t/s \in R_S$. Thus $I^{-1} \cap R_S = I^{-1}$.

Recall that S is the set $R \backslash \{Z(R) \cup Z(B)\}$. As B varies so does S. The next theorem characterizes Prüfer rings of the form $R(+)B$ for all possible choices of B; that is, for all possible choices of S.

Theorem 25.11 Let $\mathcal{U}(R)$ denote the set of units of the ring R.

(1) If $S = \mathcal{U}(R)$, then $R(+)B_S$ is its own total quotient ring and hence is a Prüfer ring.

(2) If $S = R \backslash Z(R) \supset \mathcal{U}(R)$, then $R(+)B_S$ is a Prüfer ring if and only if R is a Prüfer ring. Moreover, $R(+)B_S$ is the smallest Prüfer ring in $T(R(+)B)$ containing $R(+)B$.

(3) If $R \setminus Z(R) \supseteq S \supset \mathcal{U}(R)$, then the following are equivalent:

  (a) $R(+)B_S$ is a Prüfer ring.

  (b) For each finitely generated regular ideal $I(+)B_S$ of $R(+)B_S$, I is invertible as an ideal of R.

  (c) If $I \cap S \neq \emptyset$ and I is a finitely generated ideal, then I is invertible as an ideal of R.

PROOF: (1): If (r, m) is a regular element in $R(+)B_S$, then r is a unit in R. By Theorem 25.1(6), (r, m) is a unit in $R(+)B_S$.

(3): By Theorem 25.9, (b) and (c) are equivalent. We must show the equivalence of (a) and (b). Let $S \supset \mathcal{U}(R)$ and let I be a finitely generated ideal $I = (a_1, \ldots, a_n)$ with nonempty intersection with S. Then $I(+)B_S$ is a regular ideal of $R(+)B_S$ and is generated by the set $\{(a_1, \theta), \ldots, (a_n, \theta)\}$, Theorem 25.9. From Theorem 25.10, $(I(+)B_S)^{-1} = I^{-1}(+)B_S$. This yields $(I(+)B_S)^{-1}(I(+)B_S) = II^{-1}(+)B_S$. Therefore (a) and (b) are equivalent.

(2): The first statement is a special case of (3). For the second statement assume that $\tilde{R}(+)B_S$ is a Prüfer ring such that $R(+)B \subseteq \tilde{R}(+)B_S \subseteq R_S(+)B_S$; hence $R(+)B_S \subseteq \tilde{R}(+)B_S$.

Corollary 25.12 If R is a Prüfer ring, then $R(+)B_S$ is a Prüfer ring.

Theorem 25.13 The ring $R(+)B$ is a valuation ring if and only if R is a valuation ring of R_S and $B = B_S$.

PROOF: Assume that $R(+)B$ is a valuation ring. Then $R(+)B$ is integrally closed, hence $B = B_S$. There exists a prime ideal $P(+)B_S$ of $R(+)B_S$ such that if $(s, b) \in R_S(+)B_S \setminus R(+)B_S$, then there exists $(r, c) \in P(+)B_S$ such that $(s, b)(r, c) \in R(+)B_S \setminus$

$P(+)B_S$, Theorem 5.1. Consequently for each $s \in R_S \setminus R$, there is some $r \in P$ such that $sr \in R \setminus P$. Therefore (R, P) is a valuation pair of R_S. The proof of the converse is similar.

26. $A + B$ Rings

For a nonzero R-module B, the idealization $R(+)B$ is not reduced. To give examples of reduced rings with certain properties we need something other than idealization. To this end we define the $A+B$ rings.

Assume that D is a reduced ring and that $\mathcal{P}$ is a nonempty subset of $\operatorname{Spec} D$. If $\mathcal{A}$ is an indexing set for $\mathcal{P}$, let $I = \mathcal{A} \times N$, where N is the set of natural numbers. For each $i = (\alpha, n) \in I$, let $P_i = P_\alpha$ and $D_i = D/P_i$. Let $\prod D_i$ be the complete product of $\{D_i\}_{i \in I}$ and let $B = \sum_{i \in I} D_i$ (direct). Define $\varphi : D \to \prod D_i$ by $\varphi(d) = \{d + P_i\}_{i \in I}$. If A is the image of D under φ, define $R = A + B$. Rings formed in this manner are called $A + B$ *rings.* When we say that R is an $A + B$ ring we automatically assume that $R = A + B$, where A and B are defined as above.

If R is an $A + B$ ring, then $A \cap B = (0)$. Hence each element $x \in R$ can be written uniquely as $x = a + b$, $a \in A$ and $b \in B$. If $a \neq 0$, then x has infinitely many nonzero components. An element $x \in R$ is a zero divisor if and only if x has some zero component. Obviously R is a reduced ring. For $x \in R$ we sometimes write $x = \{x_i\}_{i \in I}$. If we wish to consider the ith component of x, write $(x)_i$.

Theorem 26.1 Let R be an $A + B$ ring. Then the minimal prime ideals of R that do not contain B are of the form $M_i = \{r \in R : (r)_i = 0\}$.

PROOF: Fix some i. Since the mapping of R to D_i given by $(r) \mapsto (r)_i$ is an epimorphism, $R/M_i \cong D_i = D/P_i$. Therefore each M_i is a prime ideal of R. By construction M_i does not contain

B. Choose b in B such that the ith component of b is 1 and the rest of the components are 0. Then $b(1-b) = 0$. Since $b \neq 0$, there exists a minimal prime ideal P of R such that $1 - b \in P$. For each $a \in M_i$, $a(1-b) = a$. Therefore $P = M_i$.

If P is a minimal prime ideal of R not containing B, we claim that $P = M_i$ for some i. Choose $b \in B \setminus P$; let $(b)_{i_1}, \ldots, (b)_{i_n}$ be the nonzero components of b. For $k = 1, \ldots, n$ choose $c(k) \in B$ such that

$$c(k)_i = \begin{cases} (b)_{i_k} (\neq 0) & \text{if } j = k \\ 0 & \text{otherwise} \end{cases}$$

Since $b = c(1) + \cdots + c(n)$, at least one of the $c(k) \in B \setminus P$. Define $d = c(k)$. Then $1 - d \in M_k$, $(1-d) = M_k$, and $M_k \subseteq P$. If $x \in P \setminus M_k$, then $dx \in (B \cap P) \setminus M_k$. Also $(1 - dx)dx = 0$ implies that $1 - dx \in M_k \subseteq P$. Therefore $1 \in P$, a contradiction.

More can be said concerning Theorem 26.1. Fix α and let $Q \supset P_\alpha$ be nonzero prime ideals of D (here $P_\alpha \in \mathcal{P}$). Let $i = (\alpha, n)$ for some n. Then $Q_i = \{r \in R : (r)_i \in Q/P_i\}$ is a prime ideal of $A + B$ that does not contain B and is not minimal.

Theorem 26.2 If $\mathcal{P} = \operatorname{Max} D$, if J is the Jacobson radical of D, and if R is the $A + B$ ring derived from D, then $R/B \cong D/J$. Thus the prime ideals of R containing B correspond to the prime ideals of D containing J.

PROOF: As above, let $\varphi : D \to R$ such that $\varphi(d) = \{d + P_i\}_{i \in I}$. Then $\varphi(D) = A$. Also $\varphi(d) = 0$ if and only if d is in J. Therefore $D/J \cong A$. It is clear that $R/B = (A+B)/B \cong A$.

Corollary 26.3 Use the same notation as in Theorem 26.2. If $J = (0)$, then the prime ideals of D correspond to the prime ideals of R/B.

Theorem 26.4 If $\mathcal{P} = \operatorname{Max} D$, then $R = A + B$ is its own total quotient ring.

PROOF: Let $a + b$ be an element in $A + B$ which is not a zero divisor. There is some $c \in D$ such that $\varphi(c) = a$. If c belongs to a maximal ideal of D, then a has infinitely many zero components; and hence $a + b$ is a zero divisor. Thus c is a unit in D. Let $\varphi(c^{-1}) = a'$ and note that $a'b/(a + b) \in B$. It follows that the inverse of $a + b$ is $a' - a'b/(a + b)$.

We return to a concept that we have used (but not named) earlier. Let $f : R \to S$ be a ring epimorphism with kernel K. If v is a valuation on S, then $w(r) = v(f(r))$ defines a valuation of R and $w^{-1}(\infty) \supseteq K$. The valuation w is called the *lifting of v by f*.

Let w be a valuation on R such that $w^{-1}(\infty) \supseteq K$. Then w induces a valuation on R/K given by $w'(r + K) = w(r)$. (The only part that needs attention is well-definedness; but this follows since $w^{-1}(\infty) \supseteq K$.) Let $R \xrightarrow{\beta} R/B \xrightarrow{\gamma} D/J$ where β is canonical and γ is the isomorphism from Theorem 26.2. Let $g = \gamma\beta$. Let p_i be the ith projection on $A + B$.

Theorem 26.5 If $\mathcal{P} = \operatorname{Max} D, R = A+B$, and J is the Jacobson radical of R, then the valuations on R are either the valuations on D/P_i lifted by p_i for some i, or the valuations on D/J lifted by g.

PROOF: This is straightforward from the two paragraphs preceding this result.

If v is a valuation on R, let (V_v, M_v) be its associated valuation pair.

Theorem 26.6 If $J = (0)$ and if a valuation w on D lifts by g to v on R, then:

(1) $V_v = \varphi(V_w) + B$.
(2) $M_v = \varphi(M_w) + B$.

(3) $v^{-1}(\infty) = \varphi(w^{-1}(\infty))$.

(4) M' is a regular maximal ideal of V_v if and only if $M' = \varphi(M) + B$ where M is a regular maximal ideal of V_w.

(5) R is the total quotient ring of V_v if and only if $D \subseteq T(V_w)$.

PROOF: These proofs are routine.

27. Examples

This section contains examples that will help delimit and illuminate the preceding theory. The examples are numbered consecutively beginning with 1. To facilitate the construction of these examples some technical results about special rings are proved. These are numbered as in the preceding sections. We adapt the notation used in Sections 25 and 26, and will use it without further reference.

Theorem 27.1 If D is a reduced ring and if $\mathcal{P} = \operatorname{Max} D$, then $R = A + B$ has Property A.

PROOF: If $r \in R$, then r is equal to some family $\{r_i\}_{i \in I}$, and r is a zero divisor of R if and only if some ith component of r is 0. Let $J = (a_1 + b_1, \ldots, a_n + b_n) \subseteq Z(R)$ be a finitely generated ideal of R, where $a_j \in A$ and $b_j \in B$. To prove that $\operatorname{Ann} J \neq (0)$, it suffices to show that there is some $i \in I$ such that $(a_j + b_j)_i = 0$ for $j = 1, \ldots, n$. There are two cases to consider.

Case 1. The set $\{a_1, \ldots, a_n\}$ generates a proper ideal of R. Let a'_i be the preimage in D of a_i. Then $(a'_1, \ldots, a'_n)$ is a proper ideal of D and is therefore contained in a maximal ideal M_λ of D. Thus for each $m \in N$ and $j = 1, \ldots, n$, we have $(a_j)_i = 0$ for $i = (\lambda, m)$. Since $b_j \in B$, b_j has only finitely many nonzero components. Thus for infinitely many $k \in N$ and $j = 1, \ldots, n$, we have $(a_j + b_j)_i = (a_j)_i = 0$ for $i = (\lambda, k)$. Fix one such

k and let $i = (\lambda, k)$. Choose an element b in B such that the ith component of b is 1 while the rest of the components are 0. Clearly b annihilates J.

Case 2. The set $\{a_1, \ldots, a_n\}$ generates R as an ideal. Choose $r_1, \ldots, r_n \in R$ such that $r_1 a_1 + \cdots + r_n a_n = 1$. If $(\sum_{j=1}^n r_j(a_j + b_j))_i \neq 0$ for all $i \in I$, then $J \not\subseteq Z(R)$ and we have a contradiction. Assume that $(\sum r_j(a_j + b_j))_i = 0$ for only finitely many i, say $\{i_1, \ldots, i_k\}$. (There can be only finitely many such i's by the restrictions placed on the a_j's and r_j's.) We claim that for some $i \in \{i_1, \ldots, i_k\}$, $(a_j + b_j)_i = 0$ for $j = 1, \ldots, n$. By way of contradiction, assume that for each $i_m \in \{i_1, \ldots, i_k\}$ there is j such that $(a_j + b_j)_{i_m} \neq 0$. Let j_m be the smallest j such that $(a_{j_m} + b_{j_m})_{i_m} \neq 0$. Define $t_j = r_j + c_j$, where

$$(c_j)_i = \begin{cases} (1 - r_j)_i & \text{if } i = i_m \text{ and } j = j_m, \quad m = 1, \ldots, k \\ (-r_j)_i & \text{if } i = i_m \text{ and } j \neq j_m, \quad m = 1, \ldots, k \\ 0 & \text{otherwise} \end{cases}$$

If $i \notin \{i_1, \ldots, i_k\}$, then $(\sum t_j(a_j + b_j))_i \neq 0$. On the other hand, if $i = i_m$ for some $m = 1, \ldots, k$, then $(\sum t_j(a_j + b_j))_i = (a_{j_m} + b_{j_m})_{i_m} \neq 0$. Thus for all $i \in I$, $(\sum t_j(a_j + b_j))_i \neq 0$. This contradicts the assumption that $J \subseteq Z(R)$. Hence for some i, $(a_j + b_j)_i = 0$ for $j = 1, \ldots, n$. As in case 1, $\operatorname{Ann} J \neq 0$. Therefore R has Property A.

There are close relationships between Property A and (a.c.), as can be seen by Corollary 2.9, Theorem 4.5, and Theorem 4.7. The first several examples explain the relationships between these two properties and the property that $\operatorname{Min} R$ is compact. Examples 1 and 2 show that, even in the reduced case, neither (a.c.) nor Property A implies the other. Examples 4 and 5 show that Theorems 4.5 and 4.7 cannot be extended to rings with nonzero nilradical.

Example 1 A reduced ring that has Property A, but not (a.c.).

PROOF: Let K be an algebraically closed field, let $D = K[X,Y]$ be the polynomial ring in two indeterminates over K, and let $\mathcal{P} = \operatorname{Max} D$. The ring $R = A + B$ has Property A by Theorem 27.1. To see that R does not have (a.c.) we show that while the image in R of the maximal ideal (X,Y) has nonzero annihilator, its annihilator is not equal to the annihilator of a principal ideal of R. Let M_λ denote the maximal ideal (X,Y) of D. For each positive integer n, define the element b_n of B by:

$$(b_n)_i = \begin{cases} 1 & \text{if } i = (\lambda, n) \\ 0 & \text{otherwise} \end{cases}$$

Furthermore, let a_x and a_y denote the respective images of X and Y in A. For each n, $b_n \in \operatorname{Ann}(a_x) \cap \operatorname{Ann}(a_y) = \operatorname{Ann}(a_x, a_y)$. Assume that $c = a + e \in R$ with $a \in A$ and $e \in B$, and that $\operatorname{Ann}(c) \supseteq \operatorname{Ann}(a_x, a_y)$. We show that this containment is proper.

Let f be the preimage of a. If $f \notin (X,Y)$, then $(a)_i \neq 0$ for each $i = (\lambda, n)$ and therefore there exists some n such that $b_n \notin \operatorname{Ann}(c) \supseteq \operatorname{Ann}(a_x, a_y)$, a contradiction. Hence $f \in (X,Y)$. There exists a maximal ideal M_β of D which contains f and at most one of X and Y. Since $(e)_i \neq 0$ for only finitely many i, there exists a specific $i = (\beta, m)$ such that $(c)_i = (a)_i = 0$. Choose an element $b \in B$ such that $(b)_i = 1$ and $(b)_j = 0$ for $j \neq i$. Then $b \in \operatorname{Ann}(c)$; but either $ba_x \neq 0$ or $ba_y \neq 0$, since X or Y does not belong to M_β. Therefore $\operatorname{Ann}(c) \supset \operatorname{Ann}(a_x, a_y)$, which proves that R does not have (a.c.).

Example 2 A reduced ring that has (a.c.), but not Property A.

PROOF: Let K and D be as in Example 1, and let $\mathcal{P}$ be the set of all nonzero principal prime ideals of D. Consider the ring $R = A + B$.

Let $f, g, h \in D$ with $\gcd\{f, g\} = h$. Denote the images of f, g, and h in A by a_f, a_g, and a_h respectively. We first show that

$\mathrm{Ann}(a_f, a_g) = \mathrm{Ann}(a_h)$. Let P_λ be an arbitrary but fixed nonzero principal prime ideal of D. Then $(f, g) \subseteq P_\lambda$ if and only if $(h) \subseteq P_\lambda$. But $(h) \subseteq P_\lambda$ if and only if $(a_h)_i = 0$ for all $i = (\lambda, n) \in I$; and similarly, $(f, g) \subseteq P_\lambda$ if and only if $(a_f)_i = 0 = (a_g)_i$ for all $i = (\lambda, n) \in I$. Consequently $\mathrm{Ann}(a_h) = \mathrm{Ann}(a_f, a_g)$. From this it follows that R does not have Property A: Take $f = X$ and $g = Y$; then $h = 1$ and we have $\mathrm{Ann}(a_X, a_Y) = \mathrm{Ann}(1) = (0)$, yet $(a_X, a_Y) \subseteq Z(R)$.

Our next goal is to prove that R has (a.c.). Let $c_1 = a_1 + b_1$ and $c_2 = a_2 + b_2$ where $a_i \in A$ and $b_i \in B$. Assume that $(c_1, c_2) \subseteq Z(R)$. If $\{a_1, a_2\}$ generates R as an ideal, then by the proof of Theorem 26.1 there are only finitely many values of i such that both $(a_1 + b_1)_i = 0$ and $(a_2 + b_2)_i = 0$. Call these values $i_1, \ldots, i_k$. Define r by: $(r)_i = 0$ for $i = i_1, \ldots, i_k$ and $(r)_i = 1$ for the rest of the i's. This element is certainly in $A + B$. Then $\mathrm{Ann}(c_1, c_2) = \mathrm{Ann}(r)$.

For the other case, assume that $\{a_1, a_2\}$ generates a proper ideal of R. Let $f_1, f_2 \in D$ be the corresponding preimages of $a_1, a_2 \in A$, let $h = \gcd\{f_1, f_2\}$ and let a_h be the image of h in A. As above $(f, g) \subseteq P_\lambda$ if and only if $(h) \subseteq P_\lambda$ for all $P_\lambda \in \mathcal{P}$; that is, $(a_1)_i = 0 = (a_2)_i$ if and only if $(a_h)_i = 0$. Because b_1 and b_2 are in B, there exist only finitely many values of i, say $i_1, \ldots, i_m$, such that at least one of $(b_1)_i$ or $(b_2)_i$ is nonzero. Of this set say that $i_1, \ldots, i_k$ $(k \leq m)$ have the property that $(a_1 + b_1)_i = 0 = (a_2 + b_2)_i$. Then for each of $i_{k+1}, \ldots, i_m$, either $(a_1 + b_1)_i \neq 0$ or $(a_2 + b_2)_i \neq 0$. Define an element $b \in B$ by

$$(b)_i = \begin{cases} -(a_h)_i & \text{if } i = i_1, \ldots, i_k \\ (1 - a_h)_i & \text{if } i = i_{k+1}, \ldots, i_m \\ 0 & \text{otherwise} \end{cases}$$

An easy argument shows that $\mathrm{Ann}(a_1 + b_1, a_2 + b_2) = \mathrm{Ann}(a_h + b)$. Therefore R satisfies (a.c.).

Example 3 A reduced ring R with (a.c.) and Property A, but where $\operatorname{Min} R$ is not compact.

PROOF: Let K be an algebraically closed field, $D = K[X]$, the polynomial ring over K, and $\mathcal{P} = \operatorname{Max} D$. By Theorem 27.1, $R = A + B$ has Property A. The proof that R has (a.c.) is the same as the corresponding proof in Example 2. (The key is that if $(f,g) \subset D$, then $(f,g) = (h)$ for some $h \in D$. Thus, for all $M_\lambda \in \mathcal{P}, (f,g) \subseteq M_\lambda$ if and only if $(h) \subseteq M_\lambda$.) Theorem 26.4 implies that R is its own total quotient ring. Also B is a non-maximal prime ideal of R (Theorem 26.2), and so R is not a von Neumann regular ring. Theorem 4.5 implies that $\operatorname{Min} R$ is not compact.

We need the following result about idealization.

Theorem 27.2 If R is a reduced ring, $\mathcal{P} = \operatorname{Max} R$ and $B = \sum_{M \in \mathcal{P}} R/M$ (direct), then $R(+)B$ has Property A.

PROOF: By Theorem 25.3, $Z(R(+)B) = Z(R) \cup Z(B) = \{(r,b) \in R(+)B : r \text{ is a nonunit of } R\}$. Hence $R(+)B$ is its own total quotient ring. Let $J = ((r_1,b_1),\ldots,(r_n,b_n))$ be a proper ideal of $R(+)B$; then $J' = (r_1,\ldots,r_n)$ is a proper ideal of R. Fix a maximal ideal M_i of R such that $J' \subseteq M_i$. Define $b \in B$ by $(b)_j = 1$ if $i = j$, or 0 otherwise. Then $(0,b) \in \operatorname{Ann} J$. Therefore $R(+)B$ has Property A.

Example 4 [cf. Theorem 4.5]. A nonreduced ring with Property A whose minimum spectrum is compact, but in which (a.c.) does not hold.

PROOF: Let K, D, and $\mathcal{P}$ be as in Example 1. Define $B = \sum_{\lambda \in \Lambda} D/M_\lambda$ (direct), where Λ is an indexing set for $\mathcal{P}$. The ring $D(+)B$ has Property A, Theorem 27.2. $\operatorname{Min} D(+)B$ is compact since it is equal to the singleton $\{0(+)B\}$. Consider the ideal $J = ((X,\theta),(Y,\theta)) \subseteq Z(D(+)B)$. The proof that there is no element $(c,b) \in D(+)B$ such that $\operatorname{Ann} J = \operatorname{Ann}((c,b))$ is much like the corresponding proof in Example 1 and will not be presented.

Example 5 [cf. Theorem 4.5]. A nonreduced ring with (a.c.) whose minimum spectrum is compact, but does not have Property A.

PROOF: Let K and D be the same as in Example 1. This time let $\mathcal{P}$ be the set of all nonzero principal prime ideals of D and let $B = \sum_{i\in I} D/P_i$, where I is an indexing set for $\mathcal{P}$. That $D(+)B$ does not have Property A is an exercise in [K, p. 63] and that $\operatorname{Min} D(+)B$ is compact follows since $(0)(+)B$ is the only minimal prime ideal of $D(+)B$.

We prove that $D(+)B$ has (a.c.). Let $((f,n),(g,m)) \subseteq Z(D(+)B)$. First consider the case when f and g are both 0. Since $n,m \in B$, $(n)_i$ and $(m)_i$ are nonzero for at most finitely many values of i. Define an element $b \in B$ by $(b)_i = 1$ anytime $(n)_i$ or $(m)_i$ is nonzero, and $(b)_i = 0$ the rest of the time. Then $\operatorname{Ann}((0,n),(0,m)) = \operatorname{Ann}((0,b))$.

If either f or g is nonzero, let $h = \gcd\{f,g\}$. Then (f,g) is a proper ideal of D and for each $P_i \in \mathcal{P}$, $(f,g) \subseteq P_i$ if and only if $(h) \subseteq P_i$. Since D is an integral domain, for each nonzero $k \in D, \operatorname{Ann}((k,\theta)) = \operatorname{Ann}((k,b))$ for every $b \in B$. It follows that $\operatorname{Ann}((h,\theta)) = \operatorname{Ann}((f,n),(g,m))$.

The next example will show that for reduced rings, the minimum spectrum of a ring R being compact does not imply that R has Property A or (a.c.). We already know this for nonreduced rings, Examples 4 and 5. Some preliminaries are needed.

Let K be a countable algebraically closed field, let I be a nonempty set, and let $K^I = \prod_{i\in I} K$. Suppose that A is a subalgebra of the K- algebra K^I. For $\varphi \in A$, define the *support* of φ as $\operatorname{supp}\varphi = \{x \in I : \varphi(x) \neq 0\}$ and the *cosupport* of φ as

$\operatorname{cosupp}\varphi = \{x \in I : \varphi(x) = 0\}$. Say that the algebra A satisfies (T) if

(i) A is a countable K-algebra.
(ii) For each nonconstant $\varphi \in A$, $\operatorname{cosupp}\varphi \neq \varnothing$.

Assume that N is the set of natural numbers.

Lemma 27.3 For each $\lambda \in A$, there is an embedding $h : A \to A_\lambda \subseteq K^{I\times N}$ such that A_λ has (T), and there exists λ_1 and λ_2 in A_λ such that $\operatorname{cosupp} h(\lambda) = \operatorname{supp}\lambda_1 \cup \operatorname{supp}\lambda_2$.

PROOF: The projection map $I \times N \to I$ defines an embedding $h : K^I \to K^{I\times N}$ given by $(h(\varphi))(x,n) = \varphi(x)$. Since $A[X,Y]$ is countable, let $\mu : A[X,Y] \to N$ be bijective. Construct elements λ_1 and λ_2 in $K^{I\times N}$ in the following way.

(a) If $(x,n) \in \operatorname{supp} h(\lambda)$, define $\lambda_1(x,n) = \lambda_2(x,n) = 0$.
(b) If $(x,n) \in \operatorname{cosupp} h(\lambda)$, consider all polynomials $f = \sum a_{ij}X^iY^j \in A[X,Y]$ such that for some i and j where $i + j \geq 1$, $x \in \operatorname{supp} a_{ij} \setminus \operatorname{supp} h(\lambda)$. In this case $\sum a_{ij}(x)X^iY^j \in K[X,Y]$ and must have a nontrivial solution (k_1,k_2). Define $\lambda_1(x,\mu(f)) = k_1$ and $\lambda_2(x,\mu(f)) = k_2$. For the remaining (x,n), let $\lambda_1(x,n) = 1 = \lambda_2(x,n)$.

Conditions (a) and (b) imply that $\operatorname{supp}\lambda_1 \cup \operatorname{supp}\lambda_2 = \operatorname{cosupp} h(\lambda)$.

We show that $A_\lambda = h(A)[\lambda_1,\lambda_2]$ satisfies (T). Since A is countable, A_λ is countable. Let $g = \sum h(a_{ij})\lambda_1^i\lambda_2^j$ be a nonconstant in A_λ. If there exist i and j such that $i + j \geq 1$ and $x \in \operatorname{supp} a_{ij} \setminus \operatorname{supp}\lambda$, then $g(x,\mu(f)) = 0$ and we are finished. Assume that no such x exists. If $(x,n) \in \operatorname{supp} h(\lambda)$, then $(x,n) \notin \operatorname{supp}\lambda_1 \cup \operatorname{supp}\lambda_2$; thus $g(x,n) = a_{00}(x)$. On the other hand if $(x,n) \in \operatorname{cosupp} h(\lambda)$, then $x \in \operatorname{cosupp} a_{ij}$ for $i + j \geq 1$. Again we have $g(x,n) = a_{00}(x)$. If a_{00} is a constant of A, then

$g = \sum h(a_{ij})\lambda_1^i\lambda_2^j$ is a constant of A_λ, a contradiction. Thus $a_{00}(x) = 0$ for some x. This finishes the proof.

Corollary 27.4 If $a, a_1, a_2 \in A$ such that $\operatorname{cosupp} a = \operatorname{supp} a_1 \cup \operatorname{supp} a_2$, then $\operatorname{cosupp} h(a) = \operatorname{supp} h(a_1) \cup \operatorname{supp} h(a_2)$.

Lemma 27.5 There exists a K-algebra $\mathcal{U}(A) \subseteq K^{I\times N^N}$ satisfying condition (T), and an injective homomorphism of K-algebras $\mathcal{U} : A \to \mathcal{U}(A)$ such that for each $\lambda \in A$, there exist λ_1 and $\lambda_2 \subseteq \mathcal{U}(A)$ such that $\operatorname{cosupp} \mathcal{U}(\lambda) = \operatorname{supp} \lambda_1 \cup \operatorname{supp} \lambda_2$.

PROOF: By Lemma 27.3, order the elements of A as $A = \{a_1, a_2, \ldots\}$, of A_{a_1} as $A_{a_1} = (b_1, b_2, \ldots\}$, of $A_{a_1a_2}$ as $A_{a_1a_2} = \{c_1, c_2, \ldots\}$ (since $a_2 \in A_{a_1}$), of $A_{a_1a_2b_1}$ as $A_{a_1a_2b_1} = \{d_1, d_2, \ldots\}$, and continue the process by the following design

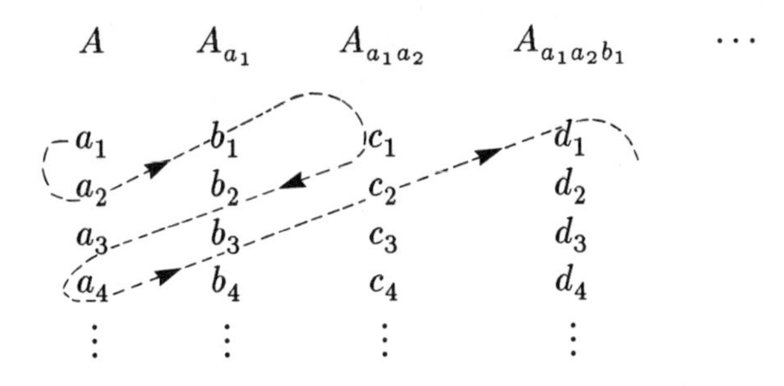

(Note: $A_{a_1a_2} = (A_{a_1})_{a_2}, A_{a_1a_2b_1} = (A_{a_1a_2})_{b_1}$, etc. Also, for example, the algebra that heads the 7th column is $A_{a_1a_2b_1c_1b_2a_3}$.) We get the following nested set of K-algebras:

$$A \subseteq A_{a_1} \subseteq A_{a_1a_2} \subseteq A_{a_1a_2b_1} \subseteq A_{a_1a_2b_1c_1} \subseteq \cdots$$

If $\mathcal{U}(A)$ denotes their union, then A can be embedded into $\mathcal{U}(A)$ and $\mathcal{U}(A)$ is the desired K-algebra.

Theorem 27.6 There is a K-algebra $\mathcal{V}(A) \subseteq K^{I\times N^N}$ satisfying condition (T), and an injective homomorphism $A \to \mathcal{V}(A)$ such that for each $\mu \in \mathcal{V}(A)$, there exist μ_1 and $\mu_2 \in \mathcal{V}(A)$ where $\operatorname{cosupp} \mu = \operatorname{supp} \mu_1 \cup \operatorname{supp} \mu_2$.

PROOF: Successive applications of Lemma 27.5 yield

$$\mathcal{U}(A) \subseteq K^{I \times N^N}$$

$$\mathcal{U}(\mathcal{U}(A)) = \mathcal{U}^2(A) \subseteq K^{I \times N^N \times N^N}$$

$$\vdots$$

Let $\mathcal{V}(A) = \cup \mathcal{U}^n(A) \subseteq K^{I \times N^N \times \cdots \times N^N \times \cdots}$

$$= K^{I \times N^{N^2}}$$

$$= K^{I \times N^N}$$

If $\mu \in \mathcal{V}(A)$, then $\mu \in \mathcal{U}^n(A)$ for some n. Hence the splitting elements μ_1 and μ_2 are in $\mathcal{U}^{n+1}(A) \subseteq \mathcal{V}(A)$. Therefore $\mathcal{V}(A)$ satisfies condition (T).

Example 6 A reduced total quotient ring R for which $\operatorname{Min} R$ is compact, but R is not a von Neumann regular ring. Thus R has neither Property A nor (a.c.).

PROOF: Choose a K-algebra $A \subseteq K^I$ satisfying (T) and construct $\mathcal{V}(A)$ as above. Since $\mathcal{V}(A)$ is a function algebra, it must be reduced. If μ is a nonconstant element in $\mathcal{V}(A)$ and if μ_1, μ_2 are splitting elements guaranteed by Theorem 27.6, then $\mu\mu_1 = \mu\mu_2 = 0$. Hence μ is a zero divisor. The constant functions are the only regular elements of $\mathcal{V}(A)$ and they are units. Therefore $T(\mathcal{V}(A)) = \mathcal{V}(A)$.

Assume that $\mathcal{V}(A)$ contains an idempotent $e \neq 0, 1$. Let a and b be distinct nonzero constants in K. Then $\beta = ae + b(1-e)$ is a nonconstant with $\operatorname{cosupp} \beta = \varnothing$, a contradiction. Thus $\mathcal{V}(A)$ is a connected ring; that is, 0 and 1 are the only idempotents.

This proves that if $\mathcal{V}(A)$ is not a field, then $\mathcal{V}(A)$ is not a von Neumann regular ring.

We use Theorem 4.3(4) to prove that $\mathrm{Min}\,\mathcal{V}(A)$ is compact. If $\mu \in \mathcal{V}(A)$, then $\mathrm{cosupp}\,\mu = \mathrm{supp}\,\mu_1 \cup \mathrm{supp}\,\mu_2$ for appropriate $\mu_i \in \mathcal{V}(A)$. Let $J = (\mu_1, \mu_2)$; then it is easy to see that $J \subseteq \mathrm{Ann}(\mu)$ and $\mathrm{Ann}(J, \mu) - (0)$.

It remains to show the existence of a K-algebra $A \subseteq K^I$ satisfying condition (T), where A is not an integral domain. Let K be a countable algebraically closed field and let $I = N$. Order K as $K = \{k_0, k_1, k_2, \ldots\}$. We may assume that $k_0 = 0$ and $k_1 = 1$.

Let 1 be the identity of K^I,

$$\varphi = (k_0, 0, k_1, 0, k_2, 0, \ldots)$$
$$\psi = (0, k_0, 0, k_1, 0, k_2, \ldots)$$

Let A be the K-algebra generated by $\{1, \varphi, \psi\}$. Then A is a ring with zero divisors satisfying condition (T). The key to seeing that A satisfies (T) is to note that if f is a nonconstant element in A, then each $k \in K$ appears as some component of f. Then $R = \mathcal{V}(A)$ is the required ring.

The next theorem summarizes the relationships between property A, (a.c.), and the compactness of min spec.

Theorem 27.7 Consider the following three conditions that a ring R may possess:

(a) R has Property A.
(b) R has (a.c.).
(c) $\mathrm{Min}\,R$ is compact.
(1) In the category of all commutative rings with identity, these conditions are independent; that is, no two of these conditions imply the third.

(2) In the category of reduced commutative rings with identity, the following hold:
   (i) (a) and $(c) \Rightarrow (b)$.
   (ii) (b) and $(c) \Rightarrow (a)$.
   (iii) (a) and $(b) \not\Rightarrow (c)$.
   (iv) $(a) \not\Rightarrow (b)$.
   (v) $(b) \not\Rightarrow (a)$.
   (vi) $(c) \not\Rightarrow (a)$ or (b).

PROOF: The proof is a consequence of Theorem 4.5 and Examples 1–6.

It was pointed out in Chapter II that there is a great difference between the structure of valuation/Prüfer domains and that of valuation/Prüfer rings. Some of these differences are obvious: for example, valuation rings need not be quasilocal. We give several examples that illustrate the bad behavior that may occur for Prüfer and valuation rings with zero divisors.

If v is a valuation on R, let (V_v, M_v), be the corresponding valuation pair.

Example 7 A valuation ring v on a total quotient ring R such that M_v is not a maximal ideal of V_v. Hence, V_v is not a Prüfer valuation ring (Theorem 6.5), and V_v is not a Marot ring (Proof of Theorem 7.7).

PROOF: Let K be an algebraically closed field and let $K[X,Y]$ be the polynomial ring over K. Let $\mathcal{N}$ be the multiplicatively closed set of elements in $K[X,Y]$ of the form $X^n g(Y)$, where $n \geq 0$ and $g(Y)$ is an element in $K[Y]$ with nonzero constant term. A typical element of $D = K[X,Y]_{\mathcal{N}}$ has the form $h = X^n f/g(Y)$, where n is some integer, $f \in K[X,Y]$, and $g(Y) \in K[Y]$ has a nonzero constant term. A valuation w can be defined on D by setting $w(h) = n$. Then $V_w = \{h \in D : n \geq 0\}$ and $M_w = XV_w$.

Note that M_w is not a maximal ideal of V_w, since $M_w \subset Q$, where $Q = (X,Y)V_w$.

The Jacobson radical of D is (0). If $\mathcal{P} = \operatorname{Max} D$, let g lift w to a valuation v on $R = A + B$ with corresponding valuation pair (V_v, M_v). Then $R = T(V_v)$; but V_v is not a Prüfer valuation ring, since $M_v = \varphi(M_w) + B \subset \varphi(Q) + B$ (Theorems 26.2 and 26.6).

Example 8 A paravaluation on a total quotient ring R that is not a valuation.

PROOF: Let K, D, and $\mathcal{P}$ be as in Example 1. Define $w(X^nY^m) = n - m\sqrt{2}$ and extend this mapping to D by defining $w(a) = 0$ if $a(\neq 0) \in K$ and $w(X^{n_1}Y^{m_1} + X^{n_2}Y^{m_2}) = \min\{n_1 - m_1\sqrt{2}, n_2 - m_2\sqrt{2}\}$. Then v is a paravaluation on D with corresponding paravaluation pair (V_w, M_w). There is no $f \in D$ such that $w(d) = \sqrt{2}$; hence v is not a valuation. Lift w to a paravaluation v on $R = A + B$.

Example 9 A Prüfer valuation ring that is not a Marot ring.

PROOF: Let Q be the field of rational numbers and let v be the rank one discrete valuation on $Q(X)$ given by $v(f/g) = \deg g - \deg f$. The valuation ring associated with v is $Q[X^{-1}]_{(X^{-1})}$. Let p be an irreducible polynomial in $Q[X]$ of degree $n > 1$. Restricting v to $Q[X][1/p]$ gives a valuation $v_0 : Q[X][1/p] \to Z \cup \{\infty\}$. (The function is clearly a paravaluation and is easily seen to be surjective.) The valuation ring of v_0 is $Q[X][1/p] \cap Q[X^{-1}]_{(X^{-1})}$. Let $B = \sum Q[X]/M$ (direct) where $M \in \operatorname{Spec} Q[X] \setminus \{(p)\}$. Define $R_0 = Q[X](+)B$ and notice that the regular elements of R_0 have the form (cp^m, b) where c is a nonzero constant and $m \geq 0$, Theorem 25.3. The total quotient ring of R_0 is $T(R_0) = Q[X]_S(+)B_S$ where $S = \{(cp^m, b) : c(\neq 0) \in Q \text{ and } m \geq 0\}$; thus $T(R_0) = Q[X][1/p](+)B_S$. Then $w : Q[X][1/p](+)B_S \to Z \cup \{\infty\}$ given by $w((a,b)) = v_0(a)$ is a valuation whose associated valuation ring is $V_w = Q[X][1/p] \cap Q[X^{-1}]_{(X^{-1})}(+)B_S$. Then $M_w = \{(a,b) \in V_w :$

$v_0(a) > 0\}$ contains the element $(1/p, \theta)$ and is regular. The set of regular elements of $T(R_0)$ is mapped by w onto nZ. The set of regular elements of M_w generates the ideal $(1/p)(+)B_S$; and in fact, this ideal is principally generated by $\{1/p(+)\theta\}$. Clearly $w((X^{n-1}/p, \theta)) = 1$, so $(X^{n-1}/p, \theta) \in M_w \setminus (1/p)(+)B_S$. This proves that V_w is a valuation ring that is not a Marot ring.

To prove that V_w is a Prüfer ring, it is sufficient to show that $D = Q[X][1/p] \cap Q[X^{-1}]_{(X^{-1})}$ is a Dedekind domain, Corollary 25.12. Since $v_0(1/p) > 0$, $Q[1/p] \subseteq D$. If L is the quotient field of D, then $Q(1/p) \subseteq L \subseteq Q(X)$; hence L is a finite algebraic extension of $Q(1/p)$. Therefore $Q[1/p]'$, the integral closure of $Q[1/p]$ in L, is a Dedekind domain. Now D is a Krull domain, so it is integrally closed. Hence D is an overring of a Dedekind domain and is therefore Dedekind.

Example 10 A discrete rank one Prüfer valuation ring that contains a nonprincipal invertible ideal.

PROOF: Consider the ring (V_w, M_w) from Example 9. Then $M_w = P(+)B_S$ where P is an appropriate prime ideal of the Dedekind domain $Q[X][1/p] \cap Q[X^{-1}]_{(X^{-1})}$. Then M_w is a regular finitely generated ideal of V_w, Theorem 25.9. Hence M_w is invertible. We have already seen that M_w is not principal.

Example 11 [cf. Theorem 8.3] A Prüfer valuation ring R that is not a Marot ring. Furthermore R contains a divisorial prime ideal P that is not the intersection of regular principal fractional ideals of R.

PROOF: Choose a Dedekind domain D that contains a nonprincipal maximal ideal M such that $M^m = (t)$ is principal for some $m > 1$. Assume that m is the minimal positive integer with this property. Let $B = \sum D/Q$ (direct), where $Q \in \operatorname{Max} D \setminus \{M\}$, and let $S = D \setminus Z(B)$. Then the idealization $R = D(+)B_S$ is a Prüfer ring. An element (x, b) is a regular nonunit in R if and only if xD is M-primary. Thus (x, b) is a regular nonunit if and only if $xD = M^{mn} = t^n D$ for some positive integer n. Since $(x, b)R = (x, \theta)R = (t^n, \theta)R, \{(t^n, \theta)R\}_{n=0}^{\infty}$ is the set of regular principal ideals of R and these regular ideals are contained in $P = M(+)B_S$; hence P is the unique maximal regular ideal of R. It is clear that P is not generated by its regular elements. Therefore (R, P) is a Prüfer valuation ring which is not Marot. Since P is invertible (it is finitely generated because M is finitely generated), $P = (P^{-1})^{-1}$. But each principal fractional ideal of R that contains P must also contain R. The proof is complete.

The following result leads to an example of a Marot ring which is not additively regular. Let I be a regular ideal of R and use $\text{Reg}(I)$ to denote the regular elements of I.

Theorem 27.8 Let R be an additively regular ring and let $J_1, \ldots, J_n, I$ be regular ideals in R. Then $I \subseteq \cup J_i$ if and only if $\text{Reg}(I) \subseteq \cup J_i$.

PROOF: One direction is obvious. We concentrate on the other. Assume there exists some $z \in I \setminus \cup P_i$. Choose a regular $x \in I \cap J_1 \cap \cdots \cap J_n$ and $u \in R$ such that $y = z + ux$ is regular. Then $\text{Reg}(I) \not\subseteq \cup J_i$, a contradiction.

Example 12 A Marot ring which is not additively regular.

PROOF: Let p be a prime number and let k be a finite field of characteristic p. Denote by R the subring $k[X^p, X^{p+1}, X^{p+2}, \ldots]$ of $k[X]$. Let $\{F_0, F_1, \ldots, F_n, G_1, G_2, \ldots\}$ be a set of irreducible polynomials of $k[X]$ such that

(a) $F_0 = X$.
(b) $F_1 = 1 + X$.
(c) $\deg F_i < 2p$ for all i.
(d) $\deg G_j \geq 2p$ for all j.
(e) No two elements of the set are associates.

(f) Each irreducible polynomial of $k[X]$ is an associate to some member of the set.

Define $K_j = k[X]/(G_j)$. Then each K_j is an R-module. Let $B = \sum K_j$ (direct) and consider the idealization $R(+)B$ of R and B. The regular elements of $R(+)B$ are the elements $(f,x) \in R(+)B$ where f is a nonzero constant in k, or $f = aX^e F_1^{e_1} \cdots F_n^{e_n}$, where $a(\neq 0) \in k$, $e \geq p$, and $e_i \geq 0$ for $i = 1, \ldots, n$.

We claim that $R(+)B$ is a Marot ring. It is sufficient to prove that if $r, s \in R(+)B$ with r regular, then the ideal $I = (r, s)$ is generated by its regular elements, Theorem 7.1. Write $r = (f,x)$ and $s = (g,y)$. By construction, $B = B_S$ where $S = R \setminus Z(B)$. Thus I is generated by (f,θ) and (g,θ), Theorem 25.9. Let h be the greatest common divisor of f and g in $k[X]$. Then $h = fF + gG$ for some $F, G \in k[X]$. Thus $(hX^p, \theta) \in I$. It follows that (hX^{2p}, θ), (hX^{2p+1}, θ), $(hX^{2p+2}, \theta), \ldots \in I$. Write $g = hG'$, $G' = a_t X^t + \cdots + a_u X^u \in k[X]$ where $a_t \neq 0 \neq a_u$, and $t \leq u$. If $t \geq 2p$, then I is generated by the regular elements $\{(f,\theta), (hX^t, \theta), \ldots, (hX^u, \theta)\}$. Assume that $t < 2p$. By the division algorithm $G' = HX^p + G''$ where $H, G'' \in k[X]$ and $\deg G'' < p < 2p$. Then $g = HhX^p + hG''$ and $(hG'', \theta) \in I$. Write $G'' = aX^k F_1^{i_1} \cdots F_n^{i_n}$ where $k < p$. Since h divides f, $h = bX^t F_1^{j_1} \ldots F_t^{j_t}$, and since $hh'' \in k[X^p, X^{p+1}, \ldots], (hh'', \theta)$ is a regular element of $R(+)B$. Therefore I is generated by $\{(f,\theta), (hX^p, \theta), (hG'', \theta)\}$; hence $R(+)B$ is a Marot ring.

For a specific example, assume that $\operatorname{char} k = 2$ and $I = (X^2, X^3)(+)B$. If $J_1 = (X^2)(+)B$ and $J_2 = (X^3)(+)B$, then an easy argument shows that $\operatorname{Reg}(I) \subseteq J_1 \cup J_2$. Suppose that $f = X^2 + X^3 + \cdots + X^6 = X^2(1 + X + \cdots + X^4)$, then $(f,\theta) \in I$. Suppose that (f,θ) is in J_i for $i = 1$ or 2. Then $f = X^2 g$, where $g \in R$. Consequently $1 + X + \cdots + X^4 = g \in R$, and therefore $X \in R$, a contradiction. By Theorem 27.8, $R(+)B$ is not an additively regular ring.

We give an easy example of a situation where Theorem 9.1 fails if "paravaluation" is replaced by "valuation." As was pointed out in Section 9, the real problem is to decide if Theorem 9.3 is true if the Marot hypothesis is deleted. The solution of this problem is unknown.

Example 13 If K is a field, then K is not the intersection of valuation subrings of $K[X]$.

It is clear that K is integrally closed in $K[X]$. But $(K,(0))$ is not a valuation subring of $K[X]$ because $(K[X],(X)) \supset (K,(0))$ and $(X) \cap K = (0)$, Theorem 5.1. If D is a domain properly between K and $K[X]$, then D contains a monic nonconstant polynomial. Hence $K[X]$ is integral over D. This proves that $K[X]$ is the only valuation overring of K (in $K[X]$). Therefore K cannot be the intersection of valuation subrings of $K[X]$.

The next two examples are concerned with the integral closedness of $R[X]$.

Example 14 A reduced integrally closed ring R with $\operatorname{Min} R$ compact for which $R[X]$ is not integrally closed.

PROOF: Let R be the ring of Example 6. Then R is its own total quotient ring, $\operatorname{Min} R$ is compact, and R is not a von Neumann regular ring. By Theorem 13.9, $R[X]$ is not integrally closed.

Assume that R is an integrally closed reduced ring. The preceding example proves that $\operatorname{Min} R$ being compact is not sufficient for $R[X]$ to be integrally closed. Neither is it necessary; for there exists a total quotient ring R with Property A and $\operatorname{Min} R$ not compact (Example 1 and Theorem 26.4), and then Theorem 13.11 implies that $R[X]$ is integrally closed.

The following example shows that the converse of Theorem 13.11 is not valid. The construction used is similar to the $A + B$ construction. This example does not rule out the possibility of the validity of the converse of Theorem 16.10.

Example 15 A reduced integrally closed ring R such that $R[X]$ is integrally closed, but R does not have Property A.

PROOF: Let k be a field, let $D = k[X, Y]$, and let Γ be the set of irreducible polynomials in D with the following two properties.

(a) If $f_1, f_2 \in \Gamma$, then $(f_1) \neq (f_2)$.
(b) For each irreducible polynomial g in D, there is some $f \in \Gamma$ such that $(f) = (g)$.

For $f \in \Gamma$, define T_f to be $D/(f)$ and denote the integral closure of T_f by T'_f. Let $I = \Gamma \times N$, where N is the set of natural numbers. For each $i = (f, n) \in I$ let $T'_i = T'_f$. Take $S = \prod_{i \in I} T'_i$ and $B = \sum_{i \in I} T'_i$ (direct). Note that k can be considered as a subfield of S. Let $R_0 = k + B$. Fix $i \in I$ and choose a maximal ideal M_i of T'_i. Define $\mathcal{M}_i = \{b \in R_0 : (b)_i \in M_i\}$. It is easy to see that

(c) The set of maximal ideals of R_0 is $\{\mathcal{M}\}_{i \in I} \cup \{B\}$.
(d) $(R_0)_{\mathcal{M}_i} \cong (T'_i)_{M_i}$ for each $i \in I$.
(e) $(R_0)_B \cong k$.

Lemma 13.2 implies that R_0 is an integrally closed ring. Let x_f and y_f be the residue classes of X and Y in $T'_{(f,n)}$. Choose u and v in S such that $(u)_{(f,n)} = x_f$ and $(v)_{(f,n)} = y_f$ for every $(f, n) \in \Gamma \times N$.

Consider the ring $R = R_0[u, v] = k[u, v] + B[u, v] = k[u, v] + B$. (This last equality holds since each $q \in B[u, v]$ has the property that $(q)_i = 0$ for all but finitely many $i \in I$.) If $b \in R$, write $b = p_b(u, v) + q_b$ where $p_b(X, Y) \in k[X, Y]$ and $q_b \in B$. If $p_b(u, v) \notin k$, then some $f \in \Gamma$ divides $p_b(X, Y)$. Thus $(b)_i = 0$ for infinitely many $i = (f, n) \in \Gamma \times N$, whence b is a zero divisor. If $p_b(u, v) = 0$, then $p_b(X,Y)$ is divisible for each $f \in \Gamma$. Thus $p_b(X,Y) = 0$. These remarks show that

(g) If $p_b(u,v) \notin k$, then $b \in Z(R)$.
(h) $k[X,Y] \cong k[u,v]$.

We prove that

(i) R is integrally closed in $T(R)$.

Let $d = b/c$ be an element of $T(R)$ which is integral over R. Since c is a regular element, $p_c(u,v)$ is a nonzero element in k. We may as well assume that $p_c(u,v) = 1$. Then $d' = d - p_b(u,v) = (q_b - p_b(u,v)q_a)/(1+q_a) \in T(R_0)$, and d' is integral over R_0. By statements (d) and (e), in conjunction with Lemma 13.2, R_0 is integrally closed. Therefore $d' \in R_0$, and hence $d \in R$. This proves (i).

The next step is to show:

(j) R does not have Property A.

Let $J = (u,v)$ be the ideal in R generated by u and v. Choose a nonzero $b \in J$. Since J is a proper ideal in R, $b \in Z(R)$. The annihilator of u is the set of $g \in R$ such that $(g)_{(f,n)} = 0$ for each $f \in \Gamma$ where $(f) \neq (X)$, and the annihilator of v is the set of $h \in R$ such that $(h)_{(f,n)} = 0$ for each $f \in \Gamma$ with $(f) \neq (Y)$; hence $\operatorname{Ann} J = \operatorname{Ann}(u) \cap \operatorname{Ann}(v) = (0)$. Therefore Property A fails in the ring R.

To complete the proof we show that if Z is an indeterminate, then $R[Z]$ is integrally closed. Let G/F be an element in $T(R[Z])$ that is integral over $R[Z]$. By the definition of R, $F = P_F + Q_F$ and $G = P_G + Q_G$ where $P_F, P_G \in k[u,v][Z]$ and $Q_F, Q_G \in B[Z]$. Since F is a regular element in $R[Z]$, $P_F \neq 0$. Using the fact that the sum $k[u,v][Z] + B[Z]$ is direct, it is easy to see that P_G/P_F is integral over $k[u,v][Z]$. But $k[u,v] \cong k[X,Y]$, so $P_G/P_F = P \in k[u,v][Z]$. Consequently $G/F = P + Q/F$ with $Q = Q_G - PQ_F \in B[Z]$. Let $i = (f,n) \in \Gamma \times N$ such that there is a coefficient of Q whose ith component $\neq 0$. Then the ith component of Q/F is integral over $T_i'[Z]$ as an element of the quotient field of $T_i'[Z]$, and therefore is in $T_i'[Z]$. Since there are only finitely many such i, $Q/F \in B[Z]$. This proves that $G/F \in R[Z]$, as was required.

Example 16 A ring R such that $R'(X)$ is not the integral closure of $R(X)$ in $T(R)(X)$. The ring R is Noetherian, so $T(R)(X) = T(R[X])$, Theorem 16.4.

PROOF: Let k be a field, let U, V, W be indeterminates, and set $R = k[U,V,W]/(U^2) = k[u,v,w]$ where u, v and w are the canonical images of U, V, and W in R. Consider the polynomial ring $R[X]$ and take $t = u/(vX+w)$ in $T(R[X]) = T(R)(X)$. Since $t^2 = 0$, t is integral over $R(X)$. It is sufficient to prove that t does not belong to $R'(X)$.

If $t \in R'(X)$, we may write $t = g_1/f$ where $g_1 \in R'[X]$, $g_1^2 = 0, f \in R[X]$, and $c(f) = R$ (Theorem 16.1). Consequently $g_1 = (ug)/d$ where d is a regular element in R and $g \in R[X]$. It follows that $udf = u(vX+w)g$. Since $d \in R, d = d_0 + eu$ with $e \in R$ and $d_0 \in k[v,w]$. Similarly

$$f = (a_0 + e_0 u) + (a_1 + e_1 u)X + \cdots + (a_n + e_n u)X^n$$
$$g = (b_0 + c_0 u) + (b_1 + c_1 u)X + \cdots + (b_m + c_m u)X^m$$

where $e_i, c_j \in R$ and $a_i, b_j \in k[v,w]$. Let $f_0 = a_0 + a_1 X + \cdots + a_n X^n$ and $g_0 = b_0 + b_1 X + \cdots + b_m X^m$. An easy argument shows that $c(f_0) = R$. Substituting into $udf = u(vX+w)g$ yields $udf = ud_0 f_0 = u(vX+w)g_0$. Since no element of $k[v,w]$ is a zero divisor of R, we have $d_0 f_0 = (vX+w)g_0$. Clearly $k[v,w]$ is isomorphic to the polynomial ring $k[V,W]$. Hence $vX+w$ is a prime element in $k[v,w][X]$. This forces $f_0 \in (vX+w)$. Therefore $c(f_0) \subseteq (v,w)R \subset R$, a contradiction.

Example 17 A strongly Prüfer ring that does not have Property A.

PROOF: Let D be a Dedekind domain containing a nonprincipal maximal ideal $M = (x, y)$. Furthermore, assume that neither (x) nor (y) is M-primary. Let $\mathcal{P} = \operatorname{Spec} D \setminus \{M, (0)\}$. Consider the $A + B$ ring R. Denote the images of x and y in R by a_x and a_y, respectively. If $\{M, P_{\lambda_1}, \ldots, P_{\lambda_k}\}$ and $\{M, P_{\mu_1}, \ldots, P_{\mu_m}\}$ are the subsets of $\operatorname{Spec} D$ containing x and y respectively, then $\{P_{\lambda_1} \ldots, P_{\lambda_k}\}$ and $\{P_{\mu_1}, \ldots, P_{\mu_m}\}$ are nonempty disjoint subsets of $\operatorname{Spec} D$. It is clear that the R-ideal (a_x, a_y) is contained in $Z(R)$. If $z \in \operatorname{Ann}(a_x, a_y)$, then for each i, $(z)_i(a_x)_i = 0 = (z)_i(a_y)_i$. This forces each $(z)_i = 0$, which proves that R does not have Property A.

It remains to show that R is a strongly Prüfer ring. It suffices to prove that if P is a maximal ideal of R, then R_P is a valuation domain. Assume that P does not contain B. Then P is a minimal prime ideal of R, Theorem 26.1. Thus R_P is a field. On the other hand suppose that $P \supseteq B$. Since $\cap_{P \in \mathcal{P}} P = (0)$, the proof of Theorem 26.2 gives us that $R/B \cong D$. Consequently $R_P \cong D_Q$ for some prime ideal Q of R; whence R_P is a valuation domain.

Among other things, the next example shows that the converse of Theorem 18.13 does not hold.

Example 18 A reduced total quotient ring R that is not strongly Prüfer. Hence R is a Prüfer ring which is not strongly Prüfer. Furthermore, the polynomial ring $R[X]$ is integrally closed.

PROOF: We use a variation of the $A + B$ construction. Let K be an algebraically closed field, let $D = K[X, Y]$, and let $\mathcal{P}$ be the set of nonzero principal prime ideals of D. For each $i = (\lambda, n) \in \mathcal{A} \times N$, let K_i represent the quotient field of D/P_i. As usual, let A be the canonical image of D in $\prod D/P_i$ as defined at the beginning of Section 26. For this example, let $B = \sum K_i$ (direct). Consider the ring $R = A + B$. If $r = a + b$ is a regular element in R, then there exists $d \in D$ such that $(a)_i = d + P_i$ for each $i \in \mathcal{A} \times N$. This forces d to be a nonzero element in the field K. Define $s \in R$ by

$$(s)_i = \begin{cases} (a^{-1})_i & \text{when } (b)_i = 0 \\ ((a+b)^{-1})_i & \text{when } (b)_i \neq 0 \end{cases}$$

Since $(b)_i \neq 0$ for only finitely many i, the element s exists. Clearly $rs = 1$; thus R is a reduced total quotient ring. As in the proofs of Theorems 26.1 and 26.2 we see that:

(1) $D \cong R/B$.
(2) If $P \in \operatorname{Max} R$ and $P \not\supseteq B$, then P is a minimal prime ideal.

Denote the images of X and Y in R by a_x and a_y, respectively. As in the proof of Example 2, the R-ideal (a_x, a_y) consists of zero divisors, yet $\operatorname{Ann}(a_x, a_y) = (0)$.

If M is the maximal ideal of D generated by X and Y, and if φ is the canonical homomorphism of D into R, then $D_M \cong R_{\varphi(M)}$. Clearly $(a_x, a_y)R_{\varphi(M)} \cong (X, Y)D_M$ is not a principal ideal. Therefore R is not a strongly Prüfer ring.

If P is a maximal ideal of R that does not contain B, then (2) implies that R_P is a field. If P contains B, then $R_P \cong D_Q$ for an appropriate prime ideal Q of D. Hence for each $P \in \operatorname{Max} R$, R_P is an integrally closed domain. Therefore the polynomial ring $R[X]$ is integrally closed, Theorem 13.3.

We give one more example of a Prüfer ring that is not strongly Prüfer. This time the concept of idealization is used.

Example 19 A nonreduced total quotient ring R that is not strongly Prüfer.

PROOF: Consider the ring $R = D(+)B$ as given in Example 5. Then (r,b) is a regular element in $D(+)B$ if and only if r is a unit in D. Hence R is a total quotient ring.

We prove that R is not a strongly Prüfer ring by showing that if Z is an indeterminate, then $R(Z)$ is not a Prüfer ring, Theorem 18.10. By Example 5 we know that (X,θ) and (Y,θ) are zero divisors and that $\text{Ann}((X,\theta),(Y,\theta)) = (0)$. Let I be the ideal in $R(Z)$ generated by (X,θ) and (Y,θ). Since I contains the polynomial $(X,\theta)Z + (Y,\theta)$, it is regular. Assume that $R(Z)$ is a Prüfer ring. Then I is an invertible ideal and hence a cancellation ideal. Thus $I^2 = ((X,\theta)^2,(Y,\theta)^2)$ [G, p. 65]. Consequently,

$$(XY,\theta) = (f/g)(X^2,\theta) + (f'/g')(Y^2,\theta)$$

where $f, f', g, g' \in R[Z]$ and $c(g) = c(g') = R$. Clearing fractions yields

$$(XY,\theta)gg' = (X^2,\theta)fg' + (Y^2,\theta)f'g$$

The last equation forces f to be in the prime ideal $((Y)(+)B)R(Z)$ and f' to be in the prime ideal $((X)(+)B)R(Z)$. Therefore $(XY,\theta)c(gg') \subseteq (X^2,\theta)c(fg')+(Y^2,\theta)c(f'g)$, and hence $(XY,\theta) \in [(X^2)(+)B] + [(Y^2)(+)B]$. This implies that XY is in the ideal (X^2,Y^2) of D, which is impossible.

Theorem 23.2 tells us precisely when a valuation ring with chained total quotient ring is or is not a chained ring. It rests completely on whether the zero divisors of the total quotient ring are or are not contained in the ring. Easy examples abound of when the valuation ring is chained. We now construct an example to show that the last statement of Theorem 23.2 is realizable.

Example 20 A valuation ring V whose total quotient ring $T(V)$ is chained, but V is not chained.

PROOF: Let $Z_{(2)}$ be the ring of integers localized at the prime ideal (2) and let $C = \langle 2^\infty \rangle$ be the 2^∞-group; that is, the infinite cocyclic 2-group. Then C is a $Z_{(2)}$-module. Form the idealization $T = Z_{(2)}(+)C$. Clearly the ideals of T are linearly ordered and $Z(T) = (2)(+)C$; hence T is a total quotient ring. Let w be the 3-adic valuation on the rational Q. Define a valuation v on T by $v((z,n)) = w(z)$. Then $V = (Z_{(2)} \cap Z_{(3)})(+)C$ is the valuation ring for v on $T(V) = T$. Since V has two maximal ideals $(2)(+)C$ and $(3)(+)C$, V is not a chained ring.

Notes

Section 25 Idealization was introduced by Nagata in [87] and later appeared in his book [N]. Results (25.1)–(25.8), some of which appeared in [51], are from Hinkle [50]. The last results of this section, (25.9)–(25.11) are essentially in Lucas' paper [68]. However some of the details are from [66].

Section 26 Griffin says in [42] that Endo [29] attributes the $A + B$ construction to Nagata, while in [45] he states that Boisen and Larsen [16] give credit to Nagata for the construction. After looking at these two papers it is not at all clear that Endo, and Boisen and Larsen, had the $A + B$ construction in mind. At any rate Griffin considered these rings independently and was the first to make them widely accessible. The results of this section are slight variations of Griffin's results [45].

Section 27 Lucas exploited the constructions presented in Sections 25 and 26 very effectively. He is responsible for seven of the examples listed in this section. Examples 1, 2, 4, and 5, and Theorems (27.1) and (27.2) appear in [67], while Examples 3 and 8 appear in [66]. Quentel [96] proved (27.6) and gave Example 6. (The details of the proofs of (27.3), (27.4), and (27.6) are from a personal correspondence from Tom Shores.) Example 7 is in [42]. There are several other examples of valuation rings that are not Prüfer; two of the earliest were by Boisen and Larsen [16], and Gilmer [35]. Examples 9, 10, and 11 are due to D. D. Anderson and Pascual [9].

The concept of an additively regular ring was implicit in Marot's paper [75] in 1968. Also the definition of a Marot ring (which he called a ring with Property P) appeared in [75]. It was not known until 1984, when Matsuda [83] showed that the Marot property is strictly weaker than the additively regular property, that these two concepts were not equivalent. Example 12 is Matsuda's example. Matsuda actually proved much more in [83]. He defines two other properties: R satisfies Property (FU), if $\operatorname{Reg}(I) \subseteq \cup_{i=1}^{n} J_i$ implies $I \subseteq \cup_{i=1}^{n} J_i$ for each finite family of regular ideals $I, J_1, \ldots, J_n$; and R satisfies Property (U), if each regular ideal of R is a union of regular principal ideals. He shows the following:

(a) additively regular $\Rightarrow$ (FU) $\Rightarrow$ Marot

$$\Uparrow$$

(U)

(b) Marot $\not\Rightarrow$ (FU) $\not\Rightarrow$ additively regular

$$\not\Downarrow$$

(U)

(c) Neither additively regular nor (U) imply the other.

Example 15 is by Nagata and appears in [1]. Example 16 is from Akiba [2]. Examples 17 and 18 are new and appear in this book for the first time; the first is by Huckaba, and the second by Lucas. Example 19 is in Hinkle and Huckaba [51]. Example 20 is in Froeschl's paper [34].

Index of Main Results

R	=	ring
D	=	integral domain (except in 26.2, 26.4, 27.1)
T	=	total quotient ring
N.A.S.C.	=	necessary and sufficient condition(s)

2.12 0-dimensional rings have Property A

3.1 Characterizations of π-regular rings

3.2 More characterizations of π-regular rings

3.3 Characterizations of von Neumann regular rings

3.5 N.A.S.C. for R to be embedded into a 0-dimensional ring

3.6 If (0) has a finite primary decomposition, then R can be embedded into a 0-dimensional ring

4.2 R reduced $\Rightarrow \operatorname{Min} R \subseteq \{M \cap R : M \in \operatorname{Spec} Q(R)\}$

4.3 Equivalent conditions for $\operatorname{Min} R$ to be compact in a reduced ring

4.5 $T(R)$ is von Neumann regular iff $\operatorname{Min} R$ is compact with either Property A or (a.c.)

4.6 For a reduced ring R, $\operatorname{Min} R$ is compact iff $T(R[X])$ is von Neumann regular

4.7 Equivalence of Property A, (a.c.), and von Neumann regular in coherent reduced rings

5.1 Equivalent conditions for R to be a valuation ring

5.2 For $P \in \operatorname{Spec} R$, there exists a valuation pair (V, M) that dominates P

5.3 When $T \setminus R$ is multiplicatively closed

5.5 Paravaluation iff dominated polynomial property

6.1 An R-module S of $T(R)$ as intersections of localizations

6.2 Equivalent conditions for R to be Prüfer

6.5 N.A.S.C. for (R, P) to be a Prüfer valuation pair

7.1 Characterizations of Marot rings

7.2 Noetherian $\Rightarrow$ few zero divisors $\Rightarrow$ additively regular $\Rightarrow$ Marot

7.3 Overrings of Marot rings are Marot

7.4 $T(R)$ is 0-dimensional $\Rightarrow$ R is additively regular

7.6	If R is Marot and $P \in \operatorname{Spec} R$, then $R_{(P)} = R_{[P]}$ and $PR_{(P)} = [P]R_{[P]}$
7.7	Characterizations of Prüfer valuation rings in the class of Marot rings
7.8	Overrings of Marot valuation rings are valuation rings
7.9	Characterization of Marot valuation rings in the class of Prüfer valuation rings
7.10	In a Marot ring, a regular ideal Q is prime (primary) iff Q is prime (primary) for its regular elements
8.1	Description of discrete rank one valuation rings
8.2	R is Krull $\Rightarrow R_{(S)}$ is Krull
8.3	$A \in \mathcal{F}^*(R)$ and R is Marot $\Rightarrow A_v$ is the intersection of principal fractional ideals
8.4	N.A.S.C. for $\mathcal{D}(R)$ to be a group
8.6	Krull $\Rightarrow$ completely integrally closed and ACC on divisorial ideals
8.10	Assume R is Krull and $\mathcal{H}$ is the minimal regular prime ideals of R. Then: (1) $\mathcal{H}$ = the set of maximal divisorial ideals of R; (2) for each $P \in \mathcal{H}$, $R_{(P)}$ is a discrete rank one valuation ring; (3) $R = \cap R_{(P)}$, $P \in \mathcal{H}$
8.13	Characterization of the essential valuation overrings of a Krull ring
8.16	$R[X]$ is a Krull ring iff R is a finite direct sum of Krull domains
8.17	Primary decompositions of regular nonunits of a Krull ring
9.1	The integral closure = the intersection of the paravaluation overrings
9.2	Characterizing integrally closed rings
9.3	If R is Marot, then R is integrally closed iff R is the intersection of its valuation overrings
10.1	Integral closure of a Noetherian ring is Krull
10.3	The approximation theorem for Krull rings

13.10	(R, M) is quasilocal and reduced, $t = g(X)/f(X) \in T(R[X])$ is integral over $R[X]$, and f has a unit coefficient $\Rightarrow t \in R[X]$
13.11	R is integrally closed, reduced, and has Property A $\Rightarrow R[X]$ is integrally closed
14.1	Properties of $R(X)$
14.2	$R(X)$ is a Marot ring with Property A; if reduced it has (a.c.)
14.5	$R(X)$ is a μ-ring
14.7	The idempotents of R and $R(X)$ coincide
15.1	Equivalent conditions for $c(f)$ to be locally principal
15.3	$R(X, Y) = R(X)(Y) = R(Y)(X)$
15.4	Finitely generated locally principal ideals of $R(X)$ are principal
15.7	$\operatorname{Pic} R(X) = (0)$
16.1	If $S = \{f \in R[X] : c(f) = R\}$ then $R'(X) = R'[X]_S$, etc.
16.2	$t \in R[X]' \cap R(X) \Rightarrow$ there exists $h \in R[X]$ such that $t - h$ is nilpotent in $R(X)$
16.3	T is a reduced extension ring of $R \Rightarrow R'(X)$ is the integral closure of $R(X)$ in $T(X)$
16.4	Property A iff $T(R)(X) = T(R[X])$
16.6	R is integrally closed with Property A, $N(R)$ is finitely generated $\Rightarrow R(X)$ is integrally closed
16.7	R is integrally closed Noetherian $\Rightarrow R(X)$ is integrally closed
16.8	R is Prüfer with Property A $\Rightarrow R(X)$ is integrally closed
16.10	Let R be reduced. If R_M is integrally closed with Property A for each $M \in \operatorname{Max} R$, then $R[X]$ is integrally closed
17.3	$\operatorname{Dim} R < \infty \Rightarrow \dim R\langle X\rangle = \dim R[X] - 1$
17.4	$\operatorname{Dim} R < \infty \Rightarrow \dim R(X) = \dim R\langle X\rangle$
17.5	$\operatorname{Dim} R < \infty$ and R Noetherian $\Rightarrow \dim R(X) = \dim R\langle X\rangle = \dim R$

Bibliography

References

[B] N. Bourbaki, *Commutative Algebra*, Herman–Addison Wesley, Paris, France, 1972.

[G] R. Gilmer, *Multiplicative Ideal Theory*, Marcel Dekker, New York, 1972.

[Fo] R. Fossum, *The Divisor Class Group Of A Krull Domain*, Springer-Verlag, New York, Heidelberg, Berlin, 1973.

[FS] L. Fuchs and L. Salce, *Modules Over Valuation Domains*, Marcel Dekker, New York, 1985.

[K] I. Kaplansky, *Commutative Rings*, Allyn and Bacon, Boston, 1970.

[L] J. Lambek, *Lectures on Rings and Modules*, Blaisdell, Waltham, Mass., 1966.

[LM] M. Larsen and P. McCarthy, *Multiplicative Theory of Ideals*, Academic Press, New York and London, 1971.

[N] M. Nagata, *Local Rings*, Interscience, New York and London, 1962.

[No] D. Northcott, *Lessons on Rings, Modules and Multiplicities*, Cambridge Univ. Press, London, 1968.

[ZSI] O. Zariski and P. Samuel, *Commutative Algebra, Vol. I.*, Van Nostrand, Princeton, N. J., 1958.

[ZSII] O. Zariski and P. Samuel, *Commutative Algebra, Vol. II*, Van Nostrand, Princeton, N.J., 1960.

1. T. Akiba, Integrally-closedness of polynomial rings, Japan J. Math *6* (1980), 67–75.

2. T. Akiba, On the normality of $R(X)$, J. Math, Kyoto Univ. *20* (1980), 749–752.

3. D. D. Anderson, Multiplication ideals, Multiplication rings, and the ring $R(X)$, Can. J. Math. *28* (1976), 760–768.

4. D. D. Anderson, Some remarks on the ring $R(X)$, Comment. Math. Univ. Sancti Pauli *26* (1977), 137–140.

5. D. D. Anderson, *The Picard group of $R(X)$ is trivial*, preprint.

6. D. D. Anderson, D. F. Anderson, and R. Markanda, The rings $R(X)$ and $R\langle X\rangle$, J. Alg. *95* (1985), 96–115.

7. D. D. Anderson and J. Huckaba, The integral closure of a Noetherian ring. II, Comm. Alg. *6* (1978), 639–657.

8. D. D. Anderson and J. Pascual, Characterizing Prüfer rings via their regular ideals, Comm. Alg. *15* (1987), 1287–1295.

9. D. D. Anderson and J. Pascual, *Regular ideals in commutative rings, sublattices of regular ideals and Prüfer rings*, to appear in J. Alg.

10. M. Arapovic, Characterizations of the 0-dimensional rings, Glas. Mat. Ser. *18* (1983), 39–46.

11. M. Arapovic, The minimal 0-dimensional overrings of commutative rings, Glas. Mat. Ser. *18* (1983), 47–52.

12. M. Arapovic, On the embedding of a commutative ring into a 0-dimensional ring, Glas. Mat. Ser. *18* (1983), 53–59.

13. J. Arnold, On the ideal theory of the Kronecker function ring and the domain $D(X)$, Can. J. Math. *21* (1969), 558–563

14. J. Arnold and J. Brewer, Kronecker function rings and flat $D[X]$-modules, Proc. Amer. Math. Soc. *27* (1971), 483–485.

15. G. Bergman, Hereditary commutative rings and centres of hereditary rings, Proc. London Math. Soc., *23* (1971), 214–236.

16. M. Boisen and M. Larsen, Prüfer and valuation rings with zero divisors, Pac. J. Math. *40* (1972), 7–12.

17. M. Boisen and M. Larsen, On Prüfer rings as images of Prüfer domains, Proc. Amer. Math. Soc. *40* (1973), 87–90.

18. J. Brewer and W. Heinzer, R Noetherian implies $R\langle X\rangle$ is a Hilbert ring, J. Alg. *67* (1980), 204–209.

19. H. Butts and W. Smith, Prüfer rings, Math. Z. *95* (1967), 196–211.

20. J. Chuchel and N. Eggert, The complete quotient ring of images of semilocal Prüfer domains, Can. J. Math. *29* (1977), 914–927.

21. I. Cohen, On the structure and ideal theory of complete local rings, Trans. Amer. Math. Soc. *59* (1946), 54–106.

22. I. Connell, On the group ring, Can. J. Math. *15* (1963), 650–685.

23. S. Cox and R. Pendleton, Rings for which certain flat modules are projective, Trans. Amer. Math. Soc. *150* (1970), 139–156.

24. E. Davis, Overrings of commutative rings I. Noetherian overrings, Trans. Amer. Math. Soc. *104* (1962), 52–61.

25. E. Davis, Overrings of commutative rings. II, integrally closed overrings, Trans. Amer. Math. Soc. *110* (1964), 196–212.

26. A. Dixon, *A polynomial ring localization:* $R\{X\}$, dissertation, U. of Missouri–Columbia, 1987.

27. P. Eakin, The converse of a well known theorem of Noetherian rings, Math Annalen *177* (1968), 278–282.

28. N. Eggert, Rings whose overrings are integrally closed in their complete quotient ring, J. reine angew. Math. *282* (1976), 88–95.

29. S. Endo, On semi-hereditary rings, J. Math. Soc. Japan *13* (1961), 109–119.

30. C. Faith, The structure of valuation rings, J. Pure Appl. Alg. *31* (1984), 7–27.

31. D. Ferrand, Trivialisation du modules projectifs. La methode de Kronecker, J. Pure Appl. Alg. *24* (1982), 261–264.

32. M. Fontana and G. Mazzola, Sur les anneaux et schémas co-discrete, Atti Accad. Naz. Lincei Rend. Cl. Sci. Fis. Mat. Natur. *56* (1974), 45–51.

33. M. Fontana and G. Mazzola, Schémas discrets et topologie constructible, C. R. Acad. Sci. Paris A *279* (1974), 463–466.

34. P. Froeschl, Chained rings, Pac. J. Math. *65* (1976), 47–53.

35. R. Gilmer, On Prüfer rings, Bull. Amer. Math. Soc. *78* (1972), 223–224.

36. R. Gilmer, *Commutative Semigroups Rings*, U. of Chicago Press, Chicago, 1984.

37. R. Gilmer and W. Heinzer, On the divisors of monic polynomials over a commutative ring, Pac. J. Math. *78* (1978), 121–131.

38. R. Gilmer and J. Hoffmann, A characterization of Prüfer domains in terms of polynomials, Pac. J. Math. *60* (1975), 81–85.

39. R. Gilmer and J. Huckaba, Δ-rings, J. Alg. *28* (1974), 414–432.

40. K. Goodearl, *Von Neumann Regular Rings*, Pitman, London, 1979.

41. J. Gräter, Integral closure and valuation rings with zero-divisors, Studia Sci. Math. Hungarica *17* (1982), 457–458.

42. M. Griffin, *Valuation theory and multiplication rings*, Queen's Math Preprint No. 1970–37, Queen's Univ., Kingston, Canada, 1970.

43. M. Griffin, *Generalizing valuations to commutative rings*, Queen's Math. Preprint No. 1970–40, Queen's Univ., Kingston, Canada, 1970.

44. M. Griffin, Prüfer rings with zero divisors, J. reine angew. Math. *239/240* (1970), 55–67.

45. M. Griffin, Valuations and Prüfer rings, Can. J. Math *26* (1974), 412–429.

46. S. Guazzone, Su alcune classi di anelli noetheriani normali, Rend. Sem. Mat. Univ. Padova *37* (1967), 258–266

47. T. Gulliksen, P. Ribenboim, and T. Viswanathan, An elementary note on group rings, J. reine angew. Math. *242* (1970), 148–162.

48. B. Hardy and T. Shores, Arithmetical semigroup rings, Can. J. Math. *32* (1980), 1361–1371.

49. M. Henriksen and M. Jerison, The space of minimal prime ideals of a commutative ring, Trans. Amer. Math. Soc. *115* (1965), 110–130.

50. G. Hinkle, *The generalized Kronecker function ring and the ring* $R(X)$, dissertation, U. of Missouri–Columbia, 1975.

51. G. Hinkle and J. Huckaba, The generalized Kronecker function ring and the ring $R(X)$, J. reine angew. Math. *292* (1977), 25–36.

52. M. Hochster, The minimal prime spectrum of a commutative ring, Can. J. Math. *23* (1971), 749–758.

53. J. Huckaba, On valuation rings that contain zero divisors, Proc. Amer. Math. Soc. *40* (1973), 9–15.

54. J. Huckaba, The integral closure of a Noetherian ring, Trans. Amer. Math. Soc. *220* (1976), 159–166.

55. J. Huckaba and J. Keller, Annihilation of ideals in commutative rings, Pac. J. Math. *83* (1979), 375–379.

56. J. Huckaba and I. Papick, Quotient rings of polynomial rings, Manuscr. Math. *31* (1980), 167–196.

57. J. Huckaba and I. Papick, A localization of $R[X]$, Can. J. Math. *23* (1981), 103–115.

58. T. Hungerford, On the structure of principal ideal rings, Pac. J. Math., *25* (1968), 543–547.

59. R. Kennedy, Krull rings, Pac. J. Math. *89* (1980), 131–136.

60. J. Kist, Minimal prime ideals in commutative semigroups, Proc. London Math. Soc. *13* (1963), 31–50.

61. I. Kozheckhov, Chained semigroup rings, (Russian), Uspekhi Matem. Nauk *29* (1974), 169–170.

62. W. Krull, Allgemeine Bewertungstheorie, J. reine angew. Math. *167* (1932), 160–196.

63. M. Larsen, Harrison primes in a ring with few zero divisors, Proc. Amer. Math. Soc. *22* (1969), 111–116.

64. M. Larson, Prüfer rings of finite character, J. reine angew. Math. *247* (1971), 92–96.

65. L. LeRiche, The ring $R\langle X\rangle$, J. Alg. *67* (1980), 327–341.

66. T. Lucas, *The Annihilator Conditions: Property (A) and (A.C.)*, dissertation, U. of Missouri–Columbia, 1983.

67. T. Lucas, Two annihilator conditions: Property (A) and (A.C.), Comm. in Alg. *14* (1986), 557–580.

68. T. Lucas, Some results on Prüfer rings, Pac. J. Math. *124* (1986), 333–343.

69. S. McAdam, Going down in polynomial rings, Can. J. Math. *23* (1971), 704–711.

70. N. McCoy, *Rings and Ideals*, Math Assoc. of Amer., Buffalo, N.Y., 1948.

71. B. McDonald and W. Waterhouse, Projective modules over rings with many units, Proc. Amer. Math. Soc. *83* (1981), 455–458.

72. K. McLean, Commutative Artinian principal ideal rings, Proc. Lond. Math. Soc. *26* (1973), 249–272.

73. M. Manis, Valuations on a commutative ring, Proc. Amer. Math. Soc. *20* (1969), 193–198.

74. P. Maroscia, Sur les anneaux de dimension zéro, Atti Accad. Naz. Lincei Rend. Cl. Sci. Fis. Mat. Natur. *56* (1974), 451–459.

75. J. Marot, Extension de la notion d'anneau valuation, Dept. Math. Faculté dec Sci. de Brest (1968), 46 pp. et 39 pp. of compléments.

76. J. Marot, Anneaux héréditaires commutatifs, C. R. Acad. Sci. Paris *269* (1969), 58–61.

77. J. Matijevic, Maximal ideal transforms of Noetherian rings, Proc. Amer. Math. Soc. *54* (1976), 49–52.

78. E. Matlis, The minimal prime spectrum of a reduced ring, Ill. J. Math. *27* (1983), 353–391.

79. R. Matsuda, Kronecker function rings, Bull. Fac. Sci. Ibaraki Univ. Math. *13* (1981), 13–24.

80. R. Matsuda, On Kennedy's problems, Comm. Math. Univ. Sancti Pauli *31* (1982), 143–145.

81. R. Matsuda, On Hinkle–Huckaba question, Math. Japonica *28* (1983), 535–539.

82. R. Matsuda, On a question posed by Huckaba–Papick, Proc. Japan Acad. *59* Ser. A (1983), 21–23.

83. R. Matsuda, On Marot rings, Proc. Japan Acad. *60* Ser. A (1984), 134–137.

84. R. Matsuda, Generalizations of multiplicative ideal theory to rings with zero divisors, Bull. Fac. Sci., Ibaraki U., Ser. A. Math. *17* (1985), 49–101.

85. A. Mewborn, Some conditions on commutative semiprime rings, J. Alg. *13* (1969), 422–431.

86. M. Nagata, A general theory of algebraic geometry over Dedekind domains. I, Amer. J. Math. *78* (1956), 78–116.

87. M. Nagata, The theory of multiplicity in general local rings, Proc. Internat. Symp. Tokyo–Nikko 1955. Sci. Council of Japan, Tokyo (1956), 191-226.

88. J. Nishimura, Note on Krull domains, J. Math, Kyoto *15* (1975), 397–400.

89. J. Ohm and P. Vicknair, Monoid rings which are valuation rings, Comm. Alg. *11* (1983), 1355–1368.

90. J. P. Olivier, Anneaux absolument plat universels et epimorphismes d'anneaux, C. R. Acad. Sci., Paris Sér. A-B *266* (1968), A317-A318.

91. G. Picavet, Ultrafilters sur un espace spectral-anneaux de Baer-anneaux á spectre minimal compact. Math. Scand. *46* (1980), 23–53.

92. R. S. Pierce, Modules over commutative regular rings, Memoirs Amer. Math. Soc. *70* (1967), 112 pp.

93. N. Popescu and C. Vraciu, Sur la structure dex anneaux absolument plats commutatifs, J. Alg. *40* (1976), 364–383.

94. D. Portelli and W. Spangher, Krull rings with zero divisors, Comm. Alg. *11* (1983), 1817–1851.

95. P. Quartararo and H. Butts, Finite unions of ideals and modules, Proc. Amer. Math. Soc. *52* (1975), 91–96.

96. Y. Quentel, Sur la compacité du spectre minimal d'un anneau, Bull. Soc. Math. France *99* (1971), 265–272. Erratum, ibid. *100* (1972), 461.

97. D. Quillen, Projective modules over polynomial rings, Invent. Math *36* (1976), 167–171.

98. L. Ratliff, On prime divisors of the integral closure of a principal ideal, J. reine angew. Math. *255* (1972), 210–220.

99. L. Ratliff, Conditions for $\mathrm{Ker}(R[X] \to R[c/d])$ to have a linear base, Proc. Amer. Math. Soc. *39* (1973), 509–514.

100. R. Raphael, Rings of quotients and π-regularity, Pac. J. Math. *39* (1971), 229–233.

101. P. Samuel, La notion de place dans un anneau, Bull. Soc. Math. France, *85* (1957), 123–133.

102. T. Shores, On generalized valuation rings, Mich. Math. J. *21* (1974), 405–409.

103. H. Storrer, Epimorphismen von Kommutativen Ringen, Com. Mat. Helvetici *43* (1968), 378–401.

104. H. Tang, Gauss' lemma, Proc. Amer. Math. Soc. *35* (1972), 372–376.

105. R. Wiegand, Modules over universal regular rings, Pac. J. Math. *39* (1971), 807–819.

Index